WHAT'S IN

Denis Owen

Dr Owen broadcasts frequently on Radio 4 programmes emanating from the BBC Natural History Unit. He is a professional ecologist who worked in Africa for a number of years but has now returned to the UK and to Oxford. *What's in a Name* is based on the Radio 4 talks bearing this title which Dr Owen researched and presented.

Acknowledgements

What's in a Name was first a series of talks broadcast on BBC Radio 4: the book is considerably expanded, with more examples and anecdotes.

I am grateful to numerous friends who offered advice about names to include and exclude, also the many listeners who wrote to agree or disagree with my interpretations of names, some of whom contributed anecdotes.

I would particularly like to thank historian Alun Jones for help in reading and understanding *The Boke of St Albans* published in 1486, and my producer Anne Blair Gould for much help and encouragement.

Published by the
British Broadcasting Corporation
35 Marylebone High Street
London W1M 4AA

ISBN 0 563 20335 8

First published 1985
© Denis Owen 1985
Typeset by Phoenix Photosetting, Chatham
Printed in England by Mackays of Chatham Ltd

Contents

1 '. . . a rose by any other name
would smell as sweet . . .'   7

2 Naming the living world   13

3 Dr Blair's shoulder-knot   23

4 Classical butterflies   32

5 Katy did, Katy didn't   38

6 Rude and witty   45

7 Unnecessary, uncertain and unkind   56

8 Old Mother Shipton and her fluffy friends   64

9 Tinker, tailor, soldier, sailor   72

10 Tools of the trade   82

11 What shall we call it?   91

12 Camberwell remembered   97

13 Magpie names   108

14 Another way to immortality   120

15 The companies of beasts and fowls   130

   Author's Book-List   137

   Index   139

– 1 –
'...a rose by any other name would smell as sweet...'

WHAT'S IN A NAME! Juliet was not concerned about Romeo's name; it was Romeo that mattered. Nevertheless he had to have a name, just as a rose must be called something. Everything and everyone needs a name but to decide on the right one is a challenge – although nothing like the challenge of trying to trace the history and meaning of a name given many years ago. Names preoccupy, intrigue, annoy and frustrate; some seem exactly right; others are a disaster or appear to be just plain silly. But all are interesting and have stories to unfold. In the preface to his book *Names*, Basil Cottle ventures the view that, 'no name is utterly dull if we are willing to look into its origin . . .

some . . . are nobly mysterious . . . be ashamed of no name, however comic or ugly or fancy it may appear'. I agree, and straight away acknowledge my indebtedness to Cottle's book which although very different in character from mine is (I think) written in much the same spirit.

My aim is to take a light-hearted look at natural history and biological names, especially the common and scientific names of plants and animals, and to try and discover how they came about and what they mean. The book is based on my BBC Radio 4 series of talks *What's in a Name* but is considerably extended, with more examples and anecdotes. As in the programmes, I frequently offer my own interpretation as to what a name might mean, and because of this I would expect others to disagree with me on many points; indeed names is a subject where disagreement flourishes, especially over common or dialect names for plants and animals.

My earliest childhood memory is of wondering about a name. I am not sure how old I was but must have been older than most people who claim to be able to recollect events while still in the pram. I was definitely walking about and looking at things because my first memory is of finding a large moth sitting on a fence in the garden of our house in south-east London. I vividly remember wanting to possess the moth; I wanted to catch and kill it so I could keep it for ever. This feeling of wanting to own it was so pervasive that even today I shudder a little at the thought of why I wanted to be so possessive. At any rate I flicked the moth off the fence into a tin and went into the house to ask my mother how best to kill it. The moth made no attempt to escape, which puzzled me a little and pleased me a lot: all I had to do now was to kill it. My mother had no experience of killing large moths but felt that gassing might do the trick, and so we held the tin over a gas burner, turned on the gas, and waited for the moth to die. But one sniff of the gas and the moth took flight and disappeared through the open door. I think I was too shocked and disappointed to make a fuss, but the experience was one I was to have again and again as I grew older because I became an avid moth collector, and as every moth collector knows, the best specimens are always the ones that get away.

But what sort of moth had I found? What was its name and was it a rare or a common species? It still matters to me; I can picture the moth on the fence, sitting in the tin, and dashing for the open door, but for some reason I have no image of a single diagnostic marking, and so identification is impossible. But whatever it was I think the event may have been responsible for developing my lifelong interest in moths.

I eventually acquired books and learnt how to collect and identify moths; I taught myself how to pin and 'set' specimens and soon obtained a good working knowledge of the local species. I concentrated my collecting on the overgrown bombed sites of war-torn south-east London: here the vegetation was rank and uncontrolled, and absolutely ideal for moths. There were privet hawks (the first one of these I ever saw flew into a VE night celebratory bonfire), yellow underwings, old ladies, grey daggers, garden carpets and a host of others. What names they all had!

I collected butterflies, too, but at first found rather few species and began to distrust what the books said.

Where were those painted ladies the books claimed were 'common in most years' – I had never seen one; and for that matter why call a butterfly a painted lady, a stupid name if ever there was one? I now think the name a good one for a butterfly that is not only beautiful but whose occurrence is unpredictable and unreliable. In 1945, the painted ladies arrived, adding glamour to the already attractive buddleia flowers that adorned the local bombed sites. I collected half a dozen magnificent specimens and did the same in the fine summer of 1947 when they appeared in even greater numbers.

The painted lady is, I think, my favourite butterfly, not because it is the most attractive, but because it seems to turn up just about anywhere at any time, often quite unexpectedly, in very large numbers. When I lived in West Africa, painted ladies used to appear at the end of the rainy season in September, usually just after there had been a severe thunderstorm from the north. They were probably blown, or perhaps had purposefully flown from the southern edge of the Sahara to the lush greenery of the coast. They stayed about a month, and then disappeared as suddenly as they had appeared. I never did discover where they went. In Uganda, I used to see them defending the tops of small hills from other butterfly intruders; they could chase off a swallowtail twice their size. In Wyoming I once saw a spectacular southward migration; this was in September and I presumed they were on their way to warmer breeding grounds to the south. A few years ago I climbed an active volcano on the island of Sumatra and upon descending, absolutely exhausted, through the steamy rain forest, I was astonished to see a single painted lady patrolling the edge of a paddy field. I have a feeling that I have always been followed about by painted ladies and nowadays when I go to a new place I become decidedly uneasy if I fail to see one. What is it about them? How much is it the butterfly and how much the name? I wish I knew.

English common names like painted lady exist for all the other species of British butterflies and for a good many foreign ones. There are also common names for all our moths (but not for most of the foreign species), our birds, mammals, fish, reptiles, amphibians, wild flowers, shrubs and trees. Each species has a name that is generally accepted, often there are widely used alternative names, and many, many, local, dialect and folk names no longer in common use.

Smaller and less obvious living things like worms, millipedes, mites and mosses rarely have generally acceptable common names like those of the better-known and more conspicuous species. But all plants and animals, large and small, conspicuous and inconspicuous, have or should have scientific names written in Latin and derived from Greek or Latin words.

My main source of information for the origin and meaning of these names is the *Oxford English Dictionary*. My copy is the two-volume 'Compact' edition in small print which requires the use of a magnifying glass which the publishers thoughtfully provide. Often, at night, I ponder the meaning of a name I have heard or read about earlier in the day, and then as soon as morning comes I check the appropriate volume, nearly always successfully, but invariably I am sidetracked into looking up other related names. For me there is seemingly no end to looking up names in the big *Dictionary*.

Once you get started, it is not difficult to find out about the origins and meanings of natural history names. But it is compulsive and like all compulsive activities it takes time. Why not give it a try?

–2–
Naming the living world

*Probably a crab would be filled with a sense of personal outrage
if it could hear us class it without ado or apology as a
crustacean, and thus dispose of it. 'I am no such thing', it
would say, 'I am myself, myself alone'*
William James, Varieties of religious experience *(1902)*

WHY SHOULD PERFECTLY ordinary-looking plants and
animals have such cumbersome and awkward-sound-
ing scientific names? Why *Ranunculus repens* for a
buttercup or *Drosophila melanogaster* for a tiny fruit fly?
What is the point of such names, and how can we be
expected to remember them?

Ever since Aristotle there have been attempts to

classify and arrange plants and animals into some sort of order. Throughout the Middle Ages and the Renaissance, educated Europeans communicated in Latin, no matter what their mother tongue. The first names that could just about be called scientific were simply Latin words for familiar plants and animals, for example *Canis* for a dog, and *Pulex* for a flea. Different kinds of dogs or fleas were described, in Latin, according to their essential characteristics, which began to make the names more like brief descriptions than scientific names as we now know them. Even more primitive attempts at classification included making such arbitrary divisions as 'useful' or 'useless' and 'wild' or 'tame' species and, for both plants and animals, species considered as 'edible' or 'inedible'. This rather unsatisfactory state of affairs continued into the sixteenth and seventeenth centuries, but as more and more specimens were brought back from the far-flung colonial empires of the European powers, the need for a better system of classification arose.

In 1735 the Swedish botanist Linnaeus published the first edition of *Systema Naturae* in which he suggested a way of naming animals, plants and minerals. The system involved the use of several, usually descriptive, Latin names, but was too clumsy and so was abandoned. Then, in 1753 for plants, and 1758 for animals (minerals were not further considered), he invented what has since become known as the binomial or Linnean system. In this every species receives two names, one for the genus and one for the species. All names are in Latin, although these may be derived from Greek words. In some instances there may be many species in a genus, or there may be just one species to a genus. Linnaeus's system forms the basis for all existing scientific names, although, of course, many have been added since the middle of the eighteenth century.

In his delightful biography, Wilfred Blunt describes Linnaeus as the 'compleat naturalist', a man who introduced order to chaos and who believed he had been appointed by God to name and classify plants and animals. Linnaeus not only named and classified specimens sent to him by others, but travelled extensively and collected his own material. In 1741 he was elected to the professorship of medicine and botany (an unlikely combination these days) at Uppsala University and he remained there until he died in 1778. Linnaeus's house, lecture room and garden are still beautifully preserved just outside Uppsala and well worth a visit for anyone travelling in that part of Sweden.

The Linnean system is best described by way of examples. Thus, man belongs to the genus *Homo*. Only one living species, *Homo sapiens,* is recognised, despite the enormous diversity of people in the world. There is fossil evidence for the existence of other species such as *Homo neanderthalensis* and *Homo erectus*, both long extinct. Three species of white butterflies belonging to the genus *Pieris* occur in Britain; additional species are found in other parts of the world. The British species are the small white, *Pieris rapae*, the large white *Pieris brassicae*, and the green-veined white, *Pieris napi*. The three are put in the same genus because they share many structural similarities, but are classified as separate species because they do not normally interbreed.

Genera (plural for genus) are gathered together and

put into families, the next biggest unit of classification, and families into orders. Orders are placed in classes, and classes into phyla (singular phylum). Finally phyla are grouped together into kingdoms. These are the basic units of classification for all plants and animals. For some groups, additional categories have been erected, such as sub-genus, sub-family, super-class, and so on, but although these are desirable in certain circumstances, they are not essential.

The small white butterfly whose scientific name is *Pieris rapae* can now be classified as:

species: *Pieris rapae*

genus: *Pieris* (which includes the large and green-veined white)

family: Pieridae (all 'white' butterflies, some of which are not actually white in colour)

order: Lepidoptera (butterflies and moths)

class: Insecta (insects)

phylum: Arthropoda (joint-legged animals, including shrimps, centipedes and millipedes)

kingdom: Animalia (animals)

Note that generic and specific names are written in italics, and that all names except the species start with a capital letter. Even if the species is named after a person or place, the specific name still starts with a small letter. Note also that butterflies are included as animals, as are shrimps, centipedes and so on – the word animal must not be used as a synonym for mammal.

The system thus involves the use of a hierarchy into which every species of animal must fit. The same system is used to classify plants. The essence of naming and classifying is to compare and contrast, to look for similarities and differences in structure. Phyla, for example, are separated from one another by much greater structural differences than genera. The species is separated by its inability to breed with other species as well as by structural characteristics which may be obvious, as in the three white butterflies just mentioned, or virtually invisible, as in some mosquitoes.

Descriptions of new species, genera, etc., must be published and accessible to all. If a description or name is not published, then so far as the rest of the world is concerned, the plant or animal does not exist. There must also be preserved specimens, called types, which can be referred to by others. Ideally these should be housed in national museums where they are readily accessible because mistakes are common. It may be necessary to disregard a 'new' name because one already exists.

How, then, is it decided what to call a plant or animal? Let us assume that specimens have been collected and have been submitted to an acknowledged expert. After checking books and museum collections he has every reason to suppose that the specimens are of an unnamed plant or animal. How does he go about inventing a name? It has to be in Latin and he has to come up with a suitable suggestion. The easiest approach is to pick on some obvious attribute of structure or colour that separates the organism from others which are similar. There is no requirement for selecting something obvious like this; instead it can be named after its place of discovery, habitat, an aspect of its behaviour, ecology or food, or after a person, perhaps, but not necessarily its discoverer. It is equally valid to invent a fanciful name based on classical literature or mythology, although this practice is less common nowadays as fewer scientists are familiar with the classics.

Let us take a few examples. The European heron is called *Ardea cinerea*. *Ardea* is Latin for heron while *cinerea* means grey. So the name for the heron is easy to interpret. The two-spot ladybird, common in gardens in summer, has two black spots on an otherwise red back. It is called *Adalia bipunctata*. The word *bipunctata* means two spots, so in this instance pattern, not colour, is picked out for the species name. Incidentally, *melanogaster*, the species name for the fruit fly mentioned at the beginning of this chapter, means black belly, another descriptive name.

As an example of a name that recalls habitat, there is the rock thrush which is called *Monticola saxatilis*. *Monticola* is Latin and means rock-dweller, something that lives among rocks. The name *saxatilis* is also Latin and again means to live among rocks. So the scientific name for the rock thrush gives a special emphasis to the bird's habitat, as indeed does its common name. The generic name of the buttercup, *Ranunculus*, mentioned earlier, is also a reference to habitat. *Ranunculus* means little frog and in a rather imaginative way refers to the fact that buttercups often grow in damp places where little frogs are likely to be found.

The scientific name of the jay is *Garrulus glandarius*. The Latin word *garrulus* recalls the sounds these birds make; they are noisy or garrulous. The word *glandarius* is also Latin and means belonging to the acorns. Thus the jay is named because it is noisy and fond of acorns, a reference to both voice and food.

Some animals have misleading scientific names. The willow tit is called *Parus montanus*. The species name *montanus* means mountain, yet the willow tit is not typically a mountain bird. It is found commonly in lowlands, so *montanus* is rather an inaccurate but acceptable name – as indeed is willow tit because although the bird is sometimes seen in willow trees it is not especially associated with them. Exactly the same is true of its near relative, the marsh tit, *Parus palustris*. The word *palustris* means marsh, but the bird is found mainly in woods, not marshes.

The linnet is called *Acanthis cannabina* because of its favourite foods. *Acanthis* is Greek and means thistles, indicating that linnets are fond of eating thistle seeds. *Cannabis* is hemp, so the species name indicates its fondness for hemp, or cannabis, seeds. The name linnet itself is a reference to its eating habits because it likes the seeds of flax, *Linum*, a plant from which linen is made.

Some names have been coined to indicate something unpleasant about the plant or animal. The generic name *Anopheles*, which means not of use or hurtful, is entirely appropriate for a group of mosquitoes that not only bite and suck blood, but also transmit malaria. The names for a number of species of poisonous toadstools warn of their toxicity to man. Satan's or devil's boletus is called *Boletus satanas*; on the other hand the much esteemed saffron milk-cap is called *Lactarius deliciosus*.

Rather more fanciful is *Chenopodium bonus-henricus*, the scientific name for the plant commonly called Good King Henry. It is a kind of wild spinach and used to be gathered and eaten before spinach was introduced and cultivated in gardens and fields. Good Henry was apparently a German elf and the king part of the name was added later for reasons that are now obscure, and is not included in the species name which, in fact, was coined by Linnaeus.

There are many rules about naming plants and animals and on the whole they are rigidly upheld. One of the most important is that nearly always we stand by priority; in other words, the first name used is the one that sticks. Some people spend much time over this question of priority. It is easy to imagine that in different parts of the world different people independently name and publish a description of a plant or animal of exactly the same species. It then becomes necessary to decide which of the names is valid. Almost invariably the answer is the one published first. Sometimes it is discovered that the oldest name published has been overlooked and never used; in such cases it has become customary to abandon the rules of priority.

Scientists can change their minds as to where species belong in the hierarchy of classification. It is possible to place plants or animals into different genera as more knowledge of their structure becomes available. Indeed classifications are constantly being revised which can make things bewildering because of the necessity to re-learn new combinations of names. A name must not be used more than once unless it is combined with a different genus, as for example with the species name *officinalis* which has been combined with dozens of genera of edible plants. But as in the rule of priority, certain exceptions are made, as in *Oenanthe*, the generic name for water dropworts, which are plants, and for wheatears, which are birds. This name has therefore been used once in the plant and once in the animal kingdom, as has *Prunella*, the generic name for self-heal, a plant, and the dunnock, a bird.

The point about a scientific name is that it must be a unique identifier by which the organism can be recognised and talked about throughout the world. For scientific work, common names are unacceptable since they vary so much from country to country and even from place to place within a country. For animals there is a rather august international body called the International Commission on Zoological Nomenclature which issues edicts about names. The headquarters is in London. The Commission consists of twenty-six eminent zoologists from about seventeen countries and it has been considering the scientific names of animals since 1895. Its work is extremely important: about 15,000 new species of animal (most of them insects) are named every year. With this rate of discovery it is no wonder that confusion and duplication arise. For some insects it is known that the same species has been described up to twenty times, and yet only one of these names can be valid. It is the job of the Commission to decide which one and make a ruling whenever there is a dispute. The problem with plants is less acute as there are fewer species, only about 300,000 species of flowering plants, about the same number as there are beetles, just one order of insects.

After 2000 years of trying to classify and name living organisms, we now have a system which, theoretically, is foolproof. But because of the enormous variety of plant and animal life, and of the relative isolation in which biologists in many countries have to work, there is still confusion. It is the job of the plant and animal classifier to sort out this confusion and to make the system work.

–3–
Dr Blair's shoulder-knot

ONE WAY TO achieve scientific immortality is to have a plant or animal named after you. It could be a scientific name, perhaps a species or genus, or if you are lucky a family in which there are several genera and species. Or it might be a common name. What often happens is that a plant or animal is given a scientific name after the person who first found it.

Take, for example, the European birds. I reckon there are about twenty-seven species named after people. There is Leach's petrel, Bewick's swan, Montagu's harrier, Barrow's goldeneye, Cory's shearwater, Tengmalm's owl, White's thrush, and Temminck's stint, to list just some of them. All honour or commemorate a person who either discovered the bird or who was involved in its discovery. The Italian zoologist Bonelli is perhaps the most distinguished as regards the common names of European birds. He has both a warbler and an eagle named after him. Bonelli's warbler is also called *Phylloscopus bonelli*, so he is particularly well remembered.

About forty eastern North American birds are named after people, most of them naturalists and travellers known for their explorations of the natural history of the New World. Best remembered is the so-called father of American ornithology, Alexander Wilson. He has a snipe, a plover, a warbler and a petrel named after him.

You cannot give a scientific name to commemorate yourself. That is simply not allowed by the rules of nomenclature. If you discover a plant or animal you believe to be new to science and want to get yourself into the history books with the name, you have to persuade a botanist or zoologist to describe and name it after you. In practice, though, people rarely seek to have animals and plants named after them. It just happens, and has to be treated as an honour.

You do not actually have to discover the plant or animal yourself. One of the most unusual names I have been able to find is that for a small North American flycatcher. Its scientific name is *Sayornis sayi*, and its common name is Say's phoebe. So, for this bird both parts of the scientific name and the common name commemorate a man, Thomas Say, who in fact was a famous entomologist. The bird was named in the nineteenth century by Charles Bonaparte (Napoleon's nephew) who also has a bird, Bonaparte's gull, named after him.

Almost equally unusual, and certainly bizarre, is Stresemann's bush-crow, *Zavattariornis stresemanni*, from Ethiopia, named in 1938 by E. Moltoni (an Italian) for Zavattari, Professor of Zoology at Padua, and Stresemann, an eminent German ornithologist. Both names commemorate scientists and provide an intriguing Italian/German combination at a crucial time in recent history.

Then there is Bewick's swan, *Cygnus bewickii*, a regular winter visitor to Britain from its Arctic breeding grounds in the Soviet Union, named after Thomas Bewick, the eighteenth-century engraver. This particular bird demonstrates the value of scientific names because it is only the British and the French that use the common name Bewick's swan. The Dutch, Swedes and Germans have other names for it.

How about Sabine's gull, a small, white, fork-tailed gull found in the Arctic and an extremely rare vagrant to Britain? It was first shot, so it was said, by Sir Edward Sabine, a noted astronomer and physicist, who named it after his brother, Joseph. The scientific name given was *Xema sabini*. I cannot help feeling that this is rather a sly way of getting round the rules in order to have your own name attached to a new bird.

The late R.E. Moreau, in whose company I enjoyed many a bird-watching trip, spent much time in East Africa studying and collecting specimens of birds. In 1938 he described a warbler new to science which had been collected in Tanganyika (now known as Tanzania). He called the warbler *Scepomyeter winifredae* after his wife Winifred and it soon became known as Mrs Moreau's warbler, not Winifred's warbler as might have been expected and as Moreau would undoubtedly have preferred.

Montagu's harrier is a bird of prey that now occurs in Britain as a rarity. It was named after George Montagu who was known for many things, not just his harrier. In 1802, for instance, he produced the first ornithological dictionary. He was also responsible for perpetuating that ridiculous name goatsucker for the nightjar, even though Gilbert White had shown by dissection that it eats mainly moths, and certainly does not suck the milk of goats.

A bird of prey with a really delightful name is Eleonora's falcon, a summer visitor to islands in the Mediterranean, and which in appearance is a little like a very dark peregrine falcon. Its scientific name, *Falco eleonorae*, was given to commemorate Eleonora of Arborea, a fourteenth-century Sardinian princess who among many other activities was responsible for the laws protecting nesting falcons, albeit so they could be exploited for falconry. Eleonora's falcon is unusual in that it nests in colonies in late summer and feeds its young on migratory warblers and other small birds on their way south from northern Europe to Africa. After the breeding season the falcons themselves fly south across Africa and evidently spend the northern winter in Madagascar – a most extraordinary journey.

There is a particularly interesting story to tell about the butterfly called the Glanville fritillary. In Britain it occurs only on the south coast of the Isle of Wight, but on the continent it is found in many places and is not at all restricted. It is tawny and dark-chequered, and its black, spiny caterpillars feed on the leaves of plantains growing near the sea. For about two centuries entomologists have been puzzled about the identity of Glanville, and even now we cannot be absolutely sure we have the story straight. Some claimed it was named after Lady Glanville, but I think this was a mistake because writers sometimes wrote Lady when they should have written lady. Confusion also arose because there were a number of Glanvilles around who might have been interested in butterflies. Almost certainly it was named after Elizabeth (or was it Eleanor?) Glanville who in the mid-eighteenth century was extremely interested in butterflies and was probably the first lady to collect them as a hobby. The first butterfly specimens (those properly pinned and set) date only from about 1702 so it is remarkable that round about this time there was a lady walking about the countryside catching and killing butterflies. No doubt her friends were disapproving. I can imagine the comments behind her back ... 'she collects *butterflies*, don't you know. What an extraordinary thing to do, and think of her poor husband.' She not only collected butterflies but sent some of her specimens to men, who at the time were just beginning to assemble collections in order to find out which species occurred where. Another anecdote about Mrs Glanville is that some of her relations tried to contest her will under the then Act of Lunacy, arguing that only someone deprived of her senses would pursue butterflies. It is interesting that in the eighteenth century collecting butterflies was frowned upon if it was done by a lady, but not if it was done by a man. These days, of course, you are likely to meet hostility from conservationists if you are seen chasing butterflies, no matter what your gender.

The first specimens of the plant genus *Buddleia* were collected in South America in the early eighteenth century by Sir William Houstoun who suggested in an unpublished manuscript that the plants should commemorate the Reverend Adam Buddle, an Essex rector, and possibly a cousin of Mrs Glanville's. Buddle was well known as a botanist and was a contemporary of the famous naturalist John Ray. He specialised in mosses and grasses and possessed an excellent herbarium which was eventually given to Sir Hans Sloane. It later became part of the Sloane Herbarium. Buddle wrote a twelve-volume English flora which many considered more reliable than Ray's synopsis of British plants, but it was never published because of the acclaim Ray received for his work. Adam Buddle had nothing to do with the discovery of *Buddleia*. The genus was named in his honour by Linnaeus, and Sir William Houstoun's wish granted. One species, *Buddleia davidii*, was introduced to Britain from China in the late nineteenth century and has since colonised waste land and old walls in many parts of the country. Its flowers are extremely attractive to butterflies which has gained it the common name of butterfly bush, although most people call it buddleia. The specific name *davidii* is for Père David, the French missionary and explorer of China, after whom Père David's deer is named.

In more recent times there is Blair's shoulder-knot, a small moth belonging to the family Noctuidae. One night in 1951, Dr K.G. Blair caught a specimen in his light-trap at Freshwater, Isle of Wight, which he could not recognise. He killed and pinned it but still could not find it in the books, so sent the specimen to a museum where it was found to be a species new to Britain, although already well known in Europe. Dr Blair then decided it ought to have an English common name and suggested stone pinion. There were already a number of moths called pinion, and the word stone described well the rather dull, flat greyish colour of the wings. But this name was not accepted, for the moth soon became known as Blair's shoulder-knot, after its discoverer, and because of the rather indistinct shoulder-knot marking at the base of the forewings. Since its discovery in the Isle of Wight, the species has spread west to Cornwall, east to Kent and north to Leicester, and at some localities is now one of the commonest moths found in October, the month in which it is on the wing. All this has been achieved in a little over thirty years, a remarkable success story.

Dr Blair was responsible for the discovery of two other species new to Britain, Blair's wainscot in 1945, and Blair's mocha in 1946, both on the Isle of Wight. The wainscot remains a rarity but the mocha has since been found in many parts of the south of England. In addition to Dr Blair, about sixteen other people are commemorated in the common names of the larger British moths, or 'macrolepidoptera' as they are some-what misleadingly called. These include, with the year of first record in brackets, Barrett's marbled coronet (1861), Dewick's plusia (1951), Eversmann's rustic (1850), Fenn's wainscot (1864, and not to be confused with the fen wainscot, a separate species), Gregson's dart (1869), Haworth's minor (1819), Vine's rustic (1879), and Webb's wainscot (1879). Blomer's rivulet was first recorded in Durham by Captain Blomer about 1832, while Mathew's wainscot was first found in Britain by Paymaster-in-Chief G.F. Mathew in 1895. Weaver's wave (note the alliteration) was first found in North Wales in 1855, but incorrectly identified. Seven years later it was found by Greening and became known as Greening's pug. Then about 1875 the moth became known as Capper's acidalia (extraordinary name) before becoming Weaver's wave once it was realised who had made the first discovery.

There are a few animal names that at first sight seem to commemorate people, but in fact only do so once removed. The mazarine blue butterfly, common in Europe but now extinct in Britain except for the occasional vagrant, is named for its colour, mazarine, a rich, deep blue which apparently was named after Cardinal Mazarin who died in 1661, or the Duchesse de Mazarin who died in 1699. On the other hand, the name of the plant genus *Gardenia* is not because it is a popular garden plant, but instead is named after someone called Garden.

I was once sitting on the beach at Freetown in Sierra Leone when I noticed, to my amazement, a very large number of insects washed up at the high-tide line. Most were beetles or bugs of one sort or another, but there were also a few wasps and dragon-flies and, to my astonishment, one large female hawk-moth. There must have been a catastrophe, a severe storm perhaps, that had taken such a variety of species out to sea to their deaths for them to be eventually washed up on the beach. The hawk-moth was still alive and struggling in its alien environment, so I collected it and thought no more about the event. There is nothing special in finding a moth, albeit a large one, that cannot straight away be recognised to species if you live in this part of Africa.

Later I operated a moth-trap, a powerful light, in my Sierra Leone garden. I soon began to find males of what appeared to be the same species as the one stranded on the beach. I collected about a dozen and sent some of them to an expert on hawk-moths who, after careful examination, concluded they belonged to an unnamed species of the genus *Libyoclanis* (later changed to *Phylloxiphia*). He named my hawk-moth and it is now called *Phylloxiphia oweni*. Examination of specimens in collections revealed the species had first been found in Sierra Leone about seventy years earlier, but had been misidentified, and that it occurs throughout West Africa east to Katanga. Not long ago I saw the moth referred to as Owen's redwing, an appropriate name because it has brick-red hindwings. Redwing, of course, is also the name for a bird, but obviously there is no possibility of confusion.

Then, in Uganda, I was once operating a different sort of trap designed to catch small insects. Among those I caught were wasps of the family Ichneu-monidae whose larvae are parasitic on caterpillars of butterflies and moths and the larvae of many other kinds of insects. I put the specimens I obtained into alcohol to preserve them and sent the collection to Dr Henry Townes, the world's expert on this group of insects. He found many species new to science, and also about forty species of a new genus. He decided this genus should be called *Owenus*, and named the type species (the first of the forty described) as *Owenus minor*. I think there was a bit of tongue in cheek with this specific name. In the event I not only have a species of hawk-moth that bears my name, but also a whole genus of tiny parasitic wasps.

It would have been nice to have discovered some-thing rather more spectacular, a new shark or parrot, perhaps. Maybe if I keep looking something will turn up, but if it does it cannot be given the generic name *Owenus* because the name has already been used. But a shark with the species name *oweni* would be all right, provided it is combined with a different generic name.

−4−
Classical butterflies

SCIENTISTS HAVE A reputation for being methodical, logical, and precise, which, some would say, makes them just a little bit boring. But in the past a few at least had a sense of poetry and imagination, especially when it came to giving scientific names to butterflies. This, I suppose, might be because of the nature of butterflies themselves. They are dainty and ephemeral-looking, or gaudy and impressive, rather like some of the gods and goddesses of classical mythology. A glance through the scientific names of butterflies from different countries yields a rich selection of names derived from Greek and Roman sources.

This was especially brought home to me when I

studied butterflies of the genus *Charaxes* in my garden at Freetown in Sierra Leone. These butterflies are large and robust, really magnificent creatures, and included among them are some of the largest butterflies found in Africa. They are also fat-bodied, at least for a butterfly, and there is a stiffening in the forewing that gives them very tough wings. Sometimes when they are feeding from the juices of fallen fruit they buffet one another with their wings, driving away potential competitors from the fruit. They remind me of vultures at a carcass.

They fly fast and high and do not visit flowers like most butterflies. Besides the juices of fallen fruit, they like to feed from animal dung and from patches of urine. Early collections of *Charaxes* held in museums contain specimens with tiny holes in the wings caused as a consequence of them being shot. Collectors used dust shot, normally reserved for shooting small birds, to bring them down from the tree-tops.

My Sierra Leone garden was very rich in butterflies, with about 300 species altogether, compared to less than seventy in the whole of Britain. Of these, about thirty were species of *Charaxes*, and they certainly were the ones that caused the most excitement. The trouble was I could hardly ever be sure what species they were because they flew high overhead and rarely settled in the garden. I would throw a piece of ripe fruit to the ground in the hope they would come to feed from its juices. I often used field glasses to try and identify them, almost as if they were birds.

After a time I decided to build a trap. I constructed a mesh cylinder about a metre high and half a metre in width, with a wooden base, and an opening just above the base of about 5 cm, large enough to allow butterflies to enter. I baited the trap with fermented banana. I bought the most rotten and awful-looking bananas I could find, put them in a jar, and let them ferment in the sun. Then I placed the banana in a shallow tray and put it on the wooden floor of the trap. The baited trap was hung from a tree in the garden, and all I had to do was to wait. In they came, big ones, little ones, all sorts of species, including many *Charaxes*. They just could not resist the temptation of the bait, and they fed and fed with their long tongues, some becoming almost completely intoxicated. They could be picked up without difficulty and some individuals could not balance themselves and fell over when they tried to stand up. Those I trapped I marked with a small spot of coloured ink, so that I could recognise individuals if they returned, and released them. Many came back again and again, and gradually a picture built up of the species present in this part of West Africa. If I did not want to use the trap, I simply put the banana bait on a table, a sort of butterfly bird-table, and watched them.

The generic name *Charaxes* is a strange one and seems to come from Greek mythology. Charaxes was the brother of Sappho, a Lesbian poetess of the seventh century BC. I suspect this is the origin of the name, but cannot be sure, because when scientists apportion names, they do not always explain exactly why they have chosen a particular one.

I soon discovered that two of the commonest species in the garden were *Charaxes castor* and *Charaxes pollux*. In Greek mythology Castor and Pollux are the twin sons of Zeus and Leda, and were worshipped by sailors in storms, sometimes even appearing to sailors at sea in bad weather. The butterflies themselves, one might suppose, should also be twins, especially as both were named by the same person. But they do not look alike as *castor* is mainly black with a band of yellowish spots across the wings, and *pollux* is bright orange with black borders to the wings.

Many other *Charaxes* in the garden had classical names. In fact, whenever the conversation in the house turned to butterflies, you could be forgiven for supposing you were among a group of classicists. Someone might ask about a gorgeous *cynthia* that had been hanging about the fruit all day, or a nice-looking *lucretius* that had dropped in and then left in a hurry. Maybe there had been an *etheocles* (son of Oedipus), or a *brutus*, the honourable man.

One species I knew from the books and which I particularly wanted to see was *Charaxes hadrianus*. It is a big, creamy-white butterfly, with black wing-tips, and maroon shoulders which give the impression of an emperor's cloak. In mythology Hadrian was a Spaniard, and was said to have spoken Greek better than Latin, and he was not just an emperor but something of an intellectual as well. I knew from my reading that *hadrianus* was known only east of Ghana, a long way from Freetown. I very much wanted to find this butterfly in Sierra Leone because if I did it would represent a considerable extension of its known geographical range, but I thought my chances slim. One day I was talking to my friend Geoffrey Field, a keen and experienced bird-watcher, who had become interested in *Charaxes* butterflies. I think he considered them honorary birds because they are so big, and because they could be trapped and marked, just as birds can be caught and ringed. He said that next time he was going on a bird-watching trip to eastern Sierra

Leone (just a little bit nearer Ghana where *hadrianus* occurs) he would take a trap with him to see if he could find something different. This he did and brought back triumphantly a magnificent *Charaxes hadrianus*, my first and only sight of this beautiful butterfly, so aptly named after Hadrian the emperor.

Also at the banana bait were many species of brown butterflies of the family Satyridae, to which our meadow brown belongs. The name Satyridae comes from Greek mythology, the Satyrs being the spirits of woods and hills who were loosely connected with fertility rites. They are said to have been grotesque, part-human, part-animal, and full of lust and revelry. The butterfly satyrs are not especially lustful, and they are certainly not grotesque, but they do seem like the spirits of woods and hills as they flit delicately between sun and shade in tropical forests. One species I found in the Freetown garden is called the evening brown because it flies at dusk and even at night, which is unusual for a butterfly. It has the species name *leda*. Now Leda is said to have been one of the parents of Castor and Pollux, so again is a reference to mythology in the butterflies of this West African garden.

Another group in the garden were members of the genus *Precis*, so called, it is claimed, because the genus is discrete and compact, not too big, and easily recognisable. The commonest species in Africa, equivalent to our small tortoiseshell, is a butterfly called *Precis oenone*, presumably named after the Greek nymph, Oenone. Another is *Precis octavia*, which might be a reference to Octavia, wife of Mark Antony, or possibly the daughter of Claudius.

Charaxes and *Precis* belong to the family Nymphalidae, one of the biggest families of butterflies, to which the red admiral and other familiar British butterflies belong. In Greek mythology, the Nymphs were supposed to have been female personifications of natural objects like rivers, trees and mountains. They were vague, fond of music, long-lived but not immortal.

The red admiral is called *Vanessa atalanta*. Atalanta was a huntress who, so the story goes, would only marry someone who could beat her in a race. It was hard to find a husband for her because she was so fleet of foot. Her future husband gained her acceptance by beating her in a race by the simple expedient of dropping a few apples every so often as he ran along. She simply could not resist picking up the apples, lost the race, and married. It is interesting, I think, that the red admiral is an extremely fleet butterfly, and is also very fond of feeding from the juices of fallen apples.

One of the nicest classical names is *Parnassius apollo* for the apollo butterfly of alpine meadows in Europe. It is now rather rare and in some countries is protected by law. It is named after Mount Parnassus in Greece where the god Apollo is supposed to have lived. Our own adonis blue also gets its name from mythology. The butterfly occurs on chalky hillsides in southern England and is a spectacular shining blue in colour, therefore is appropriately named after that beautiful youth Adonis.

And so I am glad that at least some scientists possessed the knowledge and imagination to name attractive butterflies from Greek and Roman mythology. Somehow these names are particularly easy to remember, especially if the butterfly is seen in a beautiful and memorable setting. I shall never forget the *Charaxes* butterflies in the Freetown garden and the way they imbibed the juices of fermented banana.

−5−
Katy did, Katy didn't

I ONCE LIVED IN what had been a farmhouse in the rural Midwest of the United States and I still have a vivid memory of the natural sounds on warm summer evenings in the countryside surrounding the house. There were frogs, particularly enormous bullfrogs which positively bellow, horned owls, the whippoorwill – a kind of nightjar – but above all crickets and grasshoppers. It is these last that I miss most because except for the occasional house cricket, insect sounds at night are non-existent in the area of England where I now live.

The grasshopper, or bush-cricket as it is more properly known, that attracted my attention most on those balmy Midwestern evenings was a species called *Pterophylla camellifolia*, popularly known as the katydid. As soon as darkness fell these insects could be heard calling, 'Katy did, Katy didn't', giving between two and five pulses in each burst of what must be called a song. The song is not quite the sound of a human voice but unquestionably the bush-cricket seemed to insist that poor Katy had or had not done something a trifle naughty. In the southern states this same species calls more quickly, with up to a dozen pulses in a burst of song. Other species of katydids have similar songs, but each is distinct, and an experienced listener can identify the different species with the confidence of an ornithologist who can quickly recognise the birds by their voice. Almost all species call more quickly when it is warm than when it is cool.

The sound made by katydids is produced by stridulation – a sharp edge, the scraper, at the base of one front wing is rubbed along a file-like ridge, the file, on the adjacent part of the other front wing. The wings are raised and moved back and forth so that each stroke produces a sound. Each pulse is the number of individual strikes of the scraper on the file. Only the males produce sound, the function of which is to attract females, although in some species it is also used to give an alarm or to intimidate another male.

There are throughout the world mammals, birds, frogs and insects with names that in one way or another refer to the characteristic sound they make. Just about every language has a large vocabulary of such names; even some scientific names are indicative of sounds. The name may be an attempt to describe the sound, as in howler monkey, or may be onomatopoeia, as in cuckoo and katydid. The English language possesses a wide selection of both sorts of names, some of them well established, others decidedly local or no longer in use.

There is a theory, once popular, that human speech or language started with onomatopoeia – the imitation of natural sounds of wild animals, the rustle of the wind, running water, thunder, and so on. It was called the 'bow-wow theory' by Max Müller who in the nineteenth century was Professor of Comparative Philology at the University of Oxford. Although no longer taken particularly seriously, the theory still has its merits and attractions. After all, even today parents speaking to small children call a dog a bow-wow, a duck a quack-quack, a cow a moo-moo, and a railway engine a puff-puff, despite the fact that puff-puffs are now virtually extinct.

We also have dozens of words that aptly describe the sounds made by animals. For example, dogs bark, sheep bleat, foxes yelp, frogs croak, hens cackle, snipe drum, and owls screech. Less familiar, and hardly used nowadays, are tattling jackdaws, chanting swans, creaking geese and girning boars. Somehow, most of these words sound just right.

But it is in the actual names given to animals that we find the best attempts to describe sounds. There is a laughing gull which seems to laugh, a whistling swan which makes a whistling sound as it flies overhead, a grasshopper warbler which makes a sound like a singing grasshopper, although I think one of its dialect names, the reeler, is better because the song is very like an angler's reel being wound in. The trouble with these names is that it is difficult to decide exactly what a laugh, whistle, or grasshopper sound like. The whooper swan is better named because it makes a whooping sound but you really have to hear one before you can commit the sound to memory. The name swan itself is interesting because it comes from the Old and Middle English word meaning 'sounder', which in turn is from the Latin *sonus*, a sound. Then again, a popular alternative name for the green woodpecker is yaffle, which is virtually meaningless until you have heard the bird call. Then all becomes clear because it certainly does 'yaffle'.

The name nightjar is for the churring or jarring sound of the male. Some people think it sounds like an old-fashioned spinning-wheel. It is so distinctive the bird has received a delightful assortment of local names, among them spinner, razor grinder, scissor grinder, churr owl, jar owl, wheel churr, eve churr, night churr and, less appropriately and somewhat confusingly, screech hawk, an old Berkshire name.

In the United States there is a catbird which mews so like a cat that it is easy to be fooled into thinking the cat wants to be let in. There is also a mocking-bird which mocks or mimics many other species of songbird in a most deceptive manner, and a mourning dove which makes a sad, mourning sound, as do many other species of doves. The Americans call the long-tailed duck (a rather boring, descriptive name) the old squaw, because it makes sounds like a talkative Indian squaw.

But it is names like katydid, those that imitate the sound actually made, that are the most attractive. Among British birds there is, of course, the cuckoo, but there are many others such as curlew, kittiwake, chough, and chiff-chaff which reflect the song or call. The scientific name of the woodlark is *Lullula arborea*, which indicates it sings 'lu-lu-lu', but the specific name gives the rather misleading impression it is especially associated with trees.

The jackdaw is one of the noisiest of British birds. It used to be called chawk or chatterjack in the West of England and kae or ka wattie in Scotland, names imitative or descriptive of its loud call. The bird was known as iache dawe in the sixteenth century and became jack-daw in the seventeenth century, the hyphen being dropped more recently. But we have to be careful here because jack is also a diminutive and may not necessarily be onomatopoeia. Thus we have jack snipe, which is smaller than the common snipe, and a variety of names such as jack-bird (fieldfare), jack hawk (kestrel), jack ickle (green woodpecker), jack-in-a-bottle (long-tailed tit), and jack squealer (the Leicestershire name for the swift). Some, but not all, of these birds have larger relatives. There is also the natterjack toad which is a 'small boy' compared to the common toad, but I am not sure if the jack part of the name is imitative or diminutive.

The gaudy hoopoe is another bird whose name indicates the sound it makes in the breeding season. It is rarely heard in Britain, but in southern Europe it is one of the most pervasive sounds of the countryside in spring. Its scientific name is *Upupa epops*. *Upupa* is Latin and means hoopoe, while *epops* is Greek and also means hoopoe, so for this bird the common and both scientific names refer to its voice. However, some imagination is called for because while it is easy to make out 'hoopoe' and 'upupa' it is rather more difficult to hear 'epops'.

The chiff-chaff is one of the best known of our summer visitors, but is by no means easy to tell apart from the willow-warbler. Both are greyish-green above and greyish-white below and it is only the darker legs of the chiff-chaff that make identification possible in the field, until the song is heard. The willow-warbler has a clear warbling song while the chiff-chaff's song is a simply repeated 'chiff-chaff, chiff-chaff, chiff-chiff-chaff.' The scientific name of the chiff-chaff is *Phylloscopus collybita*. *Phylloscopus* is derived from two Greek words which when put together mean, in effect, a leaf-watcher or one with interests in leaves. This is a perfect ecological description, because chiff-chaffs spend most of the time inspecting leaves for insect food, and indeed their coloration is remarkably leaf-like. The name *collybita* is also Greek and probably means money-changer, a reference to the 'chink-chink, chink-chink, chink-chink' sound of the song which resembles coins being counted out. A French local name for the chiff-chaff is *compteur d'argent*, money counter, which is most apt. Charles Swainson in *Provincial names and folk-lore of British birds* adds chip-chop and choice-and-cheep as local names that imitate the chiff-chaff's song.

The names of many North American birds imitate the sounds they make. Some are excellent examples of onomatopoeia, others are less convincing. There is a thrush called a veery which makes a sound like 'veery', a flycatcher called a phoebe which sings 'feeby', and another flycatcher called a pewee which sings a soft 'pee-wee'. It does not require much imagination to hear a rather harsh 'chickadee' from a chickadee, a relative of our willow tit, but less impressive is the killdeer, an oversized ringed plover, which sounds a little like 'kill-deer'. Then there is the willet, a shore-bird, which makes a 'willet' sound – as do many similar birds, at least to my ear! The towhee is a kind of finch and makes a sound a little like 'tow-ee' but I like better its alternative imitative name chewink. The bobolink, said to come from Bob o' Lincoln, is a bird that flies high over open grassland in the prairie states, and is named for its voice and song.

Best of all the North American birds which have names exactly imitating their voices are three species of nightjar. The three can easily be separated by listening to them. There is the whippoorwill, the bird I remember so well in the Midwest, which makes exactly that sound when on its breeding areas in the northern swamps and woods. The poorwill, its relative, simply drops the 'whip' in its song, while the chuck-will's-widow, the southerner, sings 'chuck-will's-widow, chuck-will's-widow.' All three are elusive night birds, but their songs are unmistakable. Summer nights in the North American countryside would not be the same without them.

Africa, too, has a good selection of imitative bird names. There is the hadada, the common riverside ibis of East Africa, whose alarm is 'ha-ha, ha-ha-ha' or 'ha, da-da', and the didric cuckoo, abundant almost everywhere, which stammers 'di-di-didric'. In the south, the go-away bird, a kind of turaco, threateningly urges you to 'g'way, go-away' as well as making a sound which my African bird book describes as 'a groaning grunt as if about to be ill'. Africa is full of chatterers and babblers which seem to have defied imaginative naturalists to give them imitative names. Even that veteran ornithologist, R.E. Moreau, expressed a feeling of disgust when he described the voice of the arrow-marked babbler by saying, 'their usual intercourse sounds like the bandying of filthy language in a harsh voice.'

–6–
Rude and witty

Of all our flowers, it is probably the pansy that has the greatest variety of common names. Somehow this little plant has attracted an enormous amount of attention and generated a lot of vivid imagination. Wild pansies are weeds of fields and gardens; they pop up when the soil is disturbed, but soon lose out to stronger and more competitive species. Cultivated varieties are favourite bedding plants. There are many different kinds and each year new ones appear in the gardening catalogues. There is a seemingly endless profusion of possible colour combinations and the plant is as popular today as it was in the past when so many of its fetching and fanciful names were invented.

The word pansy is believed to be derived from the French *penser* (to think) and comes from the way the flower hangs its head, as if in thought, half hiding its face. One English common name is paunce, obviously a direct rendering of the French. Pansies certainly do hang their heads in a demure sort of way, but I think it is the hiding of the face and looking down shyly that accounts for many of the older names, some of them decidedly amorous. There are, for example, three-faces-under-a-hood (a reference to the three flower colours), and heartsease, an indication of the plant's medicinal value for easing passions of the heart. Even more suggestive and enticing are love-in-idleness, cuddle-me-to-you, jump-up-and-kiss-me, kiss-me-at-the-garden-gate, and tittle-my-fancy, but there are also rather unimaginative names like herb trinity, another reference to the three flower colours.

Lords and ladies, the common wild arum of hedge-rows, whose scientific name is *Arum maculatum*, has about a hundred names. This plant, with its long purple or pale yellow flowering spikes, is absolutely unmistakable in springtime, and the rich green, spotted or unspotted leaves and red fruits are equally conspicuous later in the year. It used to be called cuckoo-pint, and indeed still is by many people, but somehow the name lords and ladies has taken over. Why should this be so?

The word pint is from the Old English and Old German pintle, meaning penis, and refers to the long, penis-like flower spike. Cuckoo might mean that it flowers when cuckoos first arrive in spring, but might equally be derived from the Anglo-Saxon *cucu* which means lively, hence lively penis. Other names are wake-pintle and wake-robin. Wake is from the Anglo-

Saxon *cwic* and has about the same meaning as *cucu*, while robin is from the French *robinet*, a cock. Many additional local names no longer in use emphasise the phallic nature of the plant and include, cuckoo cock, dog's tausel (tassle), dog's spear, dog's cock, dog's dibble, and ram's horn. The Victorians, it seems, did not like the vulgar names in common use in the seventeenth century, and so suggested polite alternatives like lords and ladies.

Besides the phallic names, there are two other groups of common names that have been used for lords and ladies. The first draws attention to the contrasting purple or pale yellow flower spikes, among them cows and calves, angels and devils, knights and ladies, soldiers and angels, stallions and mares, Adam and Eve, soldiers and sailors, and, of course, lords and ladies. The second group describes the position of the flower spike in its sheath: parson-in-the-pulpit, preacher-in-the-pulpit, lamb-in-the-pulpit, man-in-the-pulpit, jack-in-the-box, adder's tongue, and babe-in-the-cradle.

I am also inclined to the view that lords and ladies is the long purple in Ophelia's lethal garland:

> *Of crow-flowers, daisies, and long purples*
> *That liberal shepherds give a grosser name*
> *But our cold maids do dead men's fingers call them*

although many have argued that Shakespeare had in mind the early purple orchid or even the purple loosestrife. Long purple is certainly an apt description of the flower spike, but might possibly refer to the dark purple spots on the leaves of many (but not all) plants. The early purple orchid, too, has spotted leaves and, moreover, since orchid, or orchis, is Greek for testicles and is used for the plant because the root tubers have a fanciful resemblance to testicles, there is reason for supposing that Shakespeare meant this plant rather than lords and ladies. But my guess is that liberal shepherds had a whole battery of grosser names for the phallic arum of the Elizabethan countryside.

Indeed the variety of names for lords and ladies demonstrates the difficulty in trying to trace the origin and meaning of common names for many plants and animals. Tracing the meaning of a name becomes absolutely compulsive, and whenever I think I have finished my investigations I am just as unsure about what I have discovered as when I started. All our better-known and commoner plants and animals have at least one accepted common name, and often a host of others used in different parts of the country.

Take, for example, woodlice, familiar little terrestrial crustaceans found under wood and stones wherever it is damp. It is possible that they have more common names than any other British creatures, rivalling pansies and lords and ladies. Someone collected thirty-four names from Devon alone, and for the country as a whole there are over 150. Here are some of them: coffin cutters (because of their association with wood and the belief that they are responsible for the dis-integration of coffins buried in the ground), scabby sow bugs, chizzler (wood-cutting again), footballer (because some species roll up in a ball), granny picker, loafer, parson's pig, pollydishwasher, tank (a modern name), and doodlebug (even more modern). Some, like parson's pig, are decidedly cheeky, but not as rude as the names for some of our wild flowers.

Just imagine a ramble through the seventeenth-century countryside. You might see mare's fart, priest's ballocks (it is astonishing how often priests and parsons occur in rude names), black maidenhair, naked ladies, and pissabed or shitabed, while in the garden there might be open arse, horse pistle and prick madam. Some, like cowslip and oxslip (cow- or ox- slop or shit), are still used but most were banished by the Victorians. Dandelions were called pissabeds or shitabeds because if eaten they induce diarrhoea; the belief in the dangers of picking dandelion flowers persists to this day throughout the country and in many parts of Europe. Dandelion itself is rather an odd name, coming from the French *dent de lion* (lion's tooth), possibly because of the whiteness of the cut root (as white as a lion's tooth), or the yellow of the flower (like a heraldic lion with golden teeth), or more plausibly from the jagged edges to the leaves. Another rude plant is jack-by-the-hedge, a member of the cabbage family often found growing under a hedge. This name is still used and seems to have escaped Victorian censorship, possibly because the significance of the name was not appreciated. It is called jack because of its offensive smell, like a smelly jakes, or, as it is now called, a loo or lavatory. I know an old man who still calls his lavatory the jakes.

The wheatear, a small migratory bird from Africa, also escaped the censorship of its name. When I first started bird-watching I assumed that the name wheatear referred to the ear of wheat-like markings on the side of the face, but I was wrong. Wheatear used to be wheatears, but someone cut off the 's' at the end on the grounds that you could not speak of a wheatears. But you could, because *ears* is Old English for arse, or rump. The wheatear has an obvious white rump, one way to identify it as it flies away. So, wheatear means white arse, a good descriptive name but by today's standards unquestionably vulgar. Despite its vulgar name, the wheatear was at one time much sought after as a dainty. Fuller in *The Worthies of England* writes of it as, 'being no bigger than a lark, which it equalleth in the fineness of the flesh . . . The worst is, that being only seasonable in the heat of summer, and then naturally larded with lumps of fat, it is soon subject to corrupt.' Commenting on a great lord he knew, Fuller adds he must have been a man of very weak parts, because he once saw him, at a feast, feed on chickens when there were wheatears on the table.

Podiceps, a generic name for grebes, means arse-footed and was given because of the backward position of the legs. Both the great-crested and the little grebe were called arse-foot in the days when vulgar names were more acceptable than now.

Arctic skuas chase gulls and other sea-birds, forcing them to disgorge food which with remarkable skill they catch in the air before it hits the water. For a long time it was believed that skuas frightened gulls and made them defecate. This belief has produced a variety of vulgar names, including dirt bird, skait bird, shite scouter, scouty allan, dirty allan and dung hunter and, it seems, its generic name, *Stercorarius*, which indicates it feeds on dung, and is a name that comes up again in the beetle, *Geotrupes stercorarius*, which, unlike the skua, is a true dung-feeder.

But back to the cuckoo as used in cuckoo-pint. The word occurs again and again, notably for plants that blossom in the spring at the time of arrival of the first cuckoos. We have cuckoo's bread or cuckoo's meat, old names for wood sorrel, and cuckoo flower, now called lady's smock, which, according to the sixteenth-century herbalist, John Gerard, 'flowers in the spring when the cuckoo sings pleasant notes without stammering'.

Then there is cuckoo-spit, the foam produced by nymphs of the meadow spittlebug, an insect that sucks water and nutrients from the water-conducting vessels of plants. To acquire enough nutrients a nymph must take on and dispose of copious amounts of water. As the water is excreted through the anus it is pumped up into a foam, inside which the nymph lives, protected from predatory birds and spiders. Many early naturalists thought the foam was the spit of the cuckoo, but some believed it was snake spit, frog spit, toad spit, or even the spit of witches. One enterprising botanist even tried to classify plants according to their supposed ability to produce foam. Thomas Moffet (Muffett or Mouffett), writing in 1658, disagreed and added this delightful observation: 'Also from a whitish worm in frothy dew that in May sticks to plants a certain winged green creature is bred, in form like to the smallest kind of caterpillar, first it leaps, and afterwards it flies and therefore I fit to call it Locustella, a little locust. The English call the frothy matter Woodsear, as if you would say the putrefaction of wood. The Germans call it cuckow spittle.'

Lady or lady's also occur repeatedly in flower names, as in lady's smock, just mentioned, and often, but not always, alludes to Our Lady the Virgin Mary, as it does in ladybird, an insect. Past and present names for certain wild flowers provide all the necessities for a lady's boudoir. We have her furnishings in lady's cushion and lady's bedstraw, her clothes as in lady's garters, laces, glove, and slipper, and her hair as in lady's comb, looking-glass, and tresses.

Lady's fingers and, more intimately, lady's navel, refer to parts of her body, and there is also her tipple, as in lady's nightcap, although admittedly this is stretching things a bit, as the name really refers to the structure of the flower. Not all of these are still in use, but lady's slipper, an orchid, and lady's bedstraw, an attractive, soft, fern-like plant once used to stuff mattresses, remain the most used names.

There are very few, if any, gentleman names, but several make use of man, as in old man's beard, the wild clematis of chalky soils which produces fluffy, white seeds in autumn. There is also poor man's parma-cetty, an unusual name for shepherd's purse, the familiar garden weed with purse-shaped seed pods. Parma-cetty is from the Latin *sperma-ceti*, whale's sperm, which is said to be good for bruises. This is all very well, but the poor man does not want whale's sperm for his bruises; instead he would rather have a purse containing money, the best of all possible remedies.

Another common prefix word is French, applied to anything foreign or believed to be foreign, not necessarily from France. Sometimes it is used in a slightly derogatory way, as in French sparrow grass, now called star of Bethlehem, a common wild flower of the countryside. The sprouts were reputedly sold in Bath market as asparagus. No special significance seems to be attached to the apparent alliteration between asparagus and sparrow grass. There are also French beans, cowslip, grass, honeysuckle, and lavender. French nut is the walnut, in which *wal-* is Anglo-Saxon for something foreign, as in wallflower which should be spelt with one l. There are also French sorrel and French willow, the rosebay willowherb, and the Frenchmen, a name still used for red-legged partridges and for cabbage white butterflies.

Many names for plants come from their uses in herbal medicine. Some are based on magical beliefs rather than what they can really do for an illness. I like the idea that heartsease is a remedy for passions of the heart, but I doubt if many people have ever really believed in its efficacy. In 1493, a somewhat unscrupulous but clever German with the unlikely name of Phillipus Bombastus put forward the idea of the *Doctrine of Signatures*. According to the *Doctrine* every plant is signed by God, if you happened to believe in God, but certainly signed, so that you could tell what ailment or injury to use it for. Red flowers, for example, were good for poor blood, while yellow flowers were good for jaundice. As a consequence many plants were given names that indicated their precise medicinal value.

Lungwort has spotted leaves, indicating a diseased lung, and is hence a remedy for this condition. Scabious, with its scaly-looking flower head, indicates its value in treating scabies and other skin complaints, including leprosy. Saxifrage means rock-breaker and is so called because it grows in cracks and crevices and, according to the *Doctrine*, can be used to break down and dissolve bladder stones.

A local name for figwort is kernel wort because there are kernels or root tubers which were supposed to cure scrofulous glands in the neck. Its scientific name, *Scrophularia nodosa*, alludes to the connection between the plant and the ailment it cures. Scrofulous glands are swellings in the lymphatic glands and were said to 'seize the opulent'. The condition was also called king's evil because a sufferer could be cured by a touch from a king or queen. This belief persisted from the time of Edward the Confessor to the death of Queen Anne in 1714.

Pilewort, the lesser celandine, has small tubers or balls (*pila* is Latin for ball) in the roots which are useful cures for piles. Celandine is from the Greek *chelidon*, a swallow, and on first examination it might be supposed that the name originates from the fact that the flowers appear when swallows arrive. But there is another explanation which goes back as far as Aristotle and which has persisted in one form or another over the centuries. Female swallows were supposed to use celandine to restore the sight of their young, especially, as one authority has put it, when the eyes have been deliberately put out. But who would put out the eyes of baby swallows? The seventeenth-century divine, William Coles, did not know, but cautions: 'It is known to such as have skill of nature, what wonderful cure she hath of the smallest creatures, giving them knowledge of medicine to help themselves, if haply diseases annoy them. The swallow cureth her dim eyes with Celandine.' Which means we should not ask too many questions but instead make good use of what is offered.

It was widely believed that birds need good eyesight and there are several other plants that were said to restore their sight. They include hawkweed, especially good for hawks, and eyebright, which is good for linnets as well as for people. Eyebright is still used in herbal medicines intended as eye treatments.

Ragged robin is a common plant of woodland edge and hedgerow and its name seems to come from the French *robinet déchiré*, which roughly translated means tattered cock. The botanist R.C.A. Prior, with characteristic Victorian circumlocution of the vulgar, tells us that the name comes from, 'its application, upon the *Doctrine of Signatures*, to the laceration of the organ so-called, a name suggested by its finely laciniated petals.' We are back to rude names again, and so back to lords and ladies, or cuckoo-pint which, as we have seen, can be taken to mean lively penis. According to the *Doctrine of Signatures* the flower spikes attest to its aphrodisiac properties, which is one reason why it has so many common and vulgar names.

William Coles was a firm believer in the *Doctrine*, and in *The art of simpling*, published in 1656, first warns and then reassures that, 'Though Sin and Sathan have plunged mankinde into an Ocean of Infirmities, yet the mercy of God which is over all his workes, maketh Grasse to grow upon the Mountains, and Herbes for the use of men, and hath not only stamped upon them a distinct forme, but also given them particular Signatures, whereby a man may read, even in legible characters, the use of them.'

Simpling was the gathering or study of medicinal herbs, an extremey popular activity in the seventeenth century, and which is, perhaps, undergoing a small revival in the twentieth. A person that went simpling was a simplest or simpler, not to be confused with a simpleton, while simpler's joy was the name for vervain, *Verbena officinalis*, which Prior says was given because of 'the good sale a simpler had for so highly esteemed a plant'. But even in the seventeenth century it was said: 'In our times the Art of Simpling is so farre from being rewarded, that it is grown contemptible.' This quote shows that modern values were then, as now, undermining the traditional ways of doing things and earning an honest living.

–7–
Unnecessary, uncertain and unkind

I HAVE ONLY once coined a scientific name. In the 1950s I was in Kirkcudbright (now Dumfries and Galloway) in south-west Scotland collecting Scotch argus butterflies. The Scotch argus is a dark, chocolate-brown butterfly with orange bands in the forewings and a trace of similar bands in the hindwings. In the bands are circular eye-like markings, called eye-spots, which are black with white centres. I had caught quite a number of butterflies, trying hard to get good specimens in mint condition. Butterfly collectors, like stamp collectors, like to have their specimens in absolutely perfect condition. After a while I caught one without spots, just the orange bands, and I knew instantly that this must be excessively rare, perhaps unique. Later, once the specimen had been pinned and dried, I decided I would publish a description and name for this new variety, or 'aberration' as such deviants are called by collectors.

The name that came to mind was *immaculata*, which means without spots. This seemed a good, descriptive name for the specimen I now knew to be unique. But a friend who knew more Latin than me said that *immaculata* was the wrong name. He suggested *inocellata*, which means without eye-spots, a more accurate descriptive name for the specimen, and so *inocellata* it became. Once the name and the description had been published in an entomological magazine, I gave the specimen to my friend because I knew he wanted it for his own collection, but later he presented it to the Natural History Museum in London, the most appropriate home for such a rare specimen. In some ways I wish I had never coined the name. The specimen is, after all, only a variety, probably a genetic mutant, and certainly not a new species. It is just an oddity, like a defective postage stamp, and the name I gave it is totally unnecessary.

With so many species of plants and animals, naturalists sometimes run out of ideas for names for new ones. Take the insects, with more than a million named species, and an estimated million awaiting discovery. Even in one insect family, the Ichneumonidae, there are probably more species than all the vertebrates put together, and that includes all the fish in the seas and oceans of the world. Thousands of new species of insects are described every year. It is remarkable that suitable names are found for them all, and no wonder some are wrongly named or wrongly classified.

In the rain forest of West and Central Africa there is a small white butterfly with almost transparent wings and no markings whatsoever. It flies in the shade with a slow, flapping flight, and appears to bounce up and down off vegetation. It is quite different from any other known butterfly, so different that it was first described as a moth. It is called *Pseudopontia paradoxa* and has the rather unimaginative English name of moth-like white. The scientific name seems to have arisen as follows. *Pontia* had earlier been used for a genus of white butterflies, and includes *Pontia daplidice*, the Bath white, a common south European butterfly that sometimes strays to Britain. So *Pseudopontia* means false *Pontia*, indicating that we do not really know what it is, although we suspect it is some kind of white, not a *Pontia*, but like one. The specific name *paradoxa* means it is a bit of a paradox, further emphasising uncertainty over its true identity. No doubt one of these days someone will make a careful study and determine exactly where it fits into the scheme of classification. It might not be a white at all, but I am sure it is a butterfly and not a moth.

The ashy-headed parrot-bill is an Asian bird and its scientific name is *Paradoxornis alphonsianus*. *Paradoxornis* means simply a puzzling bird. It is not a parrot, even though it has a somewhat parrot-like bill, but might be some kind of titmouse or babbler. The specific name *alphonsianus* is in honour of Professor Alphonse Milne-Edwards, the nineteenth-century French zoologist, and is a rare example of a man's first name being used in a scientific name. Surnames are commonplace and ladies' first names frequent because men named animals and plants after their wives, their lady friends and, interestingly, other men's wives. *Perdix hodgsoniae*, a partridge from Tibet, is named after Mrs Hodgson, wife of B.H. Hodgson, a resident of Nepal in the mid-nineteenth century.

But back to the *pseud-* part of *Pseudopontia*. This is a common prefix in scientific names, always denoting something false or deceptive. Thus another African butterfly has the generic name *Pseudoneptis* because it looks like but certainly is not a *Neptis*. Yet another group of African butterflies is called *Pseudacraea*. Some of the species in the group are excellent mimics of the unpalatable or toxic butterflies in the family Acraeidae. *Pseudacraea* gains protection from predatory birds by looking like unpalatable species. Indeed the coloration of some species of *Pseudacraea* is so like that of the Acraeidae that it is almost impossible to tell them apart as they flit through the African rain forests.

The bare-headed rock-fowl is a very strange bird indeed. It lives in the forests of West Africa and I have sometimes seen it following trails of driver ants. As they make their way across the forest floor, these vicious insects disturb all sorts of small creatures. The rock-fowl follows the ant trail and pounces on any unfortunate animal disturbed. It eats almost anything, including cockroaches, grasshoppers, lizards and even small snakes. The scientific name of the rock-fowl is *Picathartes gymnocephalus*. *Pica* is Latin and means magpie, and indeed the bird does have a superficial resemblance to a magpie. *Kathartes* is Greek and means cleanser or purifier. The same word, spelt *Cathartes*, is used as the generic name for the American turkey vulture which, like all vultures, can pick a carcass clean. It is included in the generic name of the rock-fowl to denote the bare head which resembles that of a vulture. The name *gymnocephalus* also means bare head, so the person who named the bird was evidently very impressed by its bare-headed appearance. But as with *Pseudopontia* we are still not sure what it is related to and exactly where it should be placed in the classification system.

The turkey has a most misleading scientific name, indeed even the common name is wrong because it comes from North America, not Turkey. It is called *Meleagris gallopavo*, a curious mixture, which requires some sorting out. *Meleagris* is from the Greek, and means a kind of guinea fowl, which it is not. *Gallus* is Latin for a cock, or chicken, which it is not, while *pavo*, also Latin, means a peacock, which it is certainly not. Evidently when the turkey was first described and named it was believed, but perhaps not seriously, to be a mixture of three different types of game bird. Incidentally, *meleagris* is the species name of a flower, the fritillary. It is called *Fritillaria meleagris*, to indicate the chequered pattern on the petals, like the plumage of a guinea fowl. And *Fritillaria* is from the Latin *fritillus*, a dice box, used with a chequered board.

There is also confusion in the naming of the water vole as *Arvicola amphibius*. *Arvicola* is from the Latin *arvum* and means a ploughed field, which presumably means that someone thought water voles lived in ploughed fields, which they do not; they are usually found in water and along river-banks, as the common name implies. The name *amphibius* is good because it means the vole is at home in water and on land.

Even the chimpanzee has a misleading and rather unfair scientific name. It is called *Pan troglodytes* and the name was given as long ago as 1779. *Troglodytes* is

Greek for burrower or cave-dweller, which is nonsense since the chimpanzee spends much of its life swinging about in trees, and certainly does not inhabit caves. Evidently when chimpanzee specimens were first brought back to Europe they were believed to be cavemen.

There is an element of fun in the naming of the laughing jackass or kookaburra, an Australian kingfisher. The story goes that an appropriate name could not be found, so it was called *Dacelo*, an anagram of *Alcedo*, the generic name of the European kingfisher.

Many plants and animals were named after the place they came from but, needless to say, mistakes were made. The name for the brown rat is *Rattus norvegicus*, but the animal has no special association with Norway. It is found throughout the world and its place of origin is not known with certainty. The stripe-bellied weasel, *Grammogale africana*, comes from Brazil, and was described as coming from Africa by mistake, but the rules of nomenclature mean that the name must stand.

I like especially Professor Scopoli's name for the little ringed plover. Scopoli was Professor of Botany at Padua in the eighteenth century and somehow became involved in naming birds, which would be unlikely to happen to a professor of botany these days. The bird is called *Charadrius dubius* and Scopoli called it *dubius* because he was not absolutely sure that it was really distinct from the closely similar ringed plover. But it is distinct, although I must admit to sometimes having trouble separating the two species in the field.

For more fun, how about the white-bellied swiftlet of Asia whose name is *Collocalia esculenta?* *Kolla* is Greek and means glue, while *kalia*, also Greek, means a dwelling, so the generic name means glue dwelling. The name *esculenta* is Latin and means good to eat; the same name has been given to many edible plants. So the white-bellied swiftlet has a scientific name which means a glue dwelling that is good to eat. It is, of course, the bird that makes a nest of saliva which once hardened is the essential ingredient of bird's nest soup.

The city of Athens is the source of all worldly wisdom, and indeed Athene was the goddess of wisdom. For centuries Athens and its people have been associated with the owl, supposedly a wise bird. The owl in question is probably the little owl which now has the name *Athene noctua*. This is all well and good, except that *noctua* is a poor choice as this particular owl is partly diurnal. *Athene diurna* might have been a better choice.

Some scientific names are decidedly unkind. There is a tern called the brown noddy found chiefly in tropical and subtropical waters. The word noddy means stupid; sometimes one person will refer to another as being a bit noddy, which presumably has something to do with nodding the head in a silly sort of way. The scientific name is *Anous stolidus*. *Anous* is Greek and means silly and *stolidus* is Latin and means dull or slow of mind. Why this unkind set of names? All because the bird is tame, easy to approach, and not afraid of people. Poor, unfortunate bird. But then what about *Homo sapiens* as the scientific name for man? The word *sapiens* is Latin and means wise and sensible, attributes which many of us would dispute. After all we have given some unnecessary, uncertain and unkind names for the other animals that share our world!

And it is not just scientific names that can be unkind. Take the wryneck, a small, woodpecker-like bird that spends the winter in Africa and arrives in Britain about the same time as the cuckoo. In various parts of the country it has been called cuckoo's mate, footman, fool, messenger, marrow (companion or friend) and leader, while when translated, a Welsh name means cuckoo's knave. Some of these names suggest that the wryneck is in some way subservient to the cuckoo. But it is certainly very difficult to see why as the two birds have totally different habits and, although they may occur in the same places, they are rarely seen together. The wryneck was at one time quite widespread in Britain but is now very rare, although still common in many parts of continental Europe.

−8−
Old Mother Shipton and her fluffy friends

THE NAMES GIVEN to plants and animals often describe colour and pattern, or some conspicuous feature of anatomy. Many birds, for example, have practical descriptive names: a blackbird is simply a black bird, a green woodpecker is a green bird that pecks wood, and a treecreeper is a bird that creeps up the trunks of trees. The name golden oriole is more interesting. It is a tautology, as both names mean golden, a good description of the bird's plumage. Whoever gave the name must have been impressed by its strikingly beautiful coloration.

When it comes to moths there seems to have been more imagination associated with the names they have received. How about a maiden's blush, a lovely name for a lovely moth, common in woods and hedgerows in summer? The blush refers to the pinkish flush on the wings of one that has just emerged from the chrysalis. Once the moth has been flying around it begins to pale and the blush disappears, a romantic version of a maiden, perhaps, and soon it looks distinctly tatty. There is a true-lover's knot, a small moth with knot-like markings on the wings, but why true-lover's is a mystery, except that knots are traditionally linked with lovers.

Rather more bizarre is the setaceous hebrew character. Setaceous means bristly, while hebrew character refers to markings on the wings which, with imagination, look a little like Hebrew script. It would be possible to translate the name as meaning bad-tempered Jew, but that is not what was meant. The lackey moth is a little easier. It is a small, brown and rather undistinguished-looking moth, but the caterpillar is furry and brightly coloured, reminding someone years ago of the gaudy and extravagant colours of a footman's or lackey's uniform.

Over two thousand species of moths are found in Britain and each has a common as well as a scientific name. These names delight and puzzle, and are often incomprehensible. You might, depending on where you live, find the alchymist, the uncertain, the stranger, the cousin german, the anomalous, and the delicate. Searching around there might be an exile, a phoenix, a saxon, and a sprawler. In most cases we can only guess at what the inventor of the name had in mind. Cousin german, for example, might have something to do with the use of the word german to mean a degree of relationship or kinship, as in Germans, an

English name for a group of related North European peoples, but not a name used by the Germans themselves. I doubt if the name of the moth has anything directly to do with Germany, but it is dull-coloured and ordinary-looking, easily confused with similar related species, which is perhaps a clue to the curious name.

Some moths are named after the food-plant of the caterpillar. There is a turnip moth (which sometimes feeds on turnip stems, but also the stems of just about any low-growing plant), a cabbage moth (which feeds on the leaves of cabbage and many other garden plants), a water betony, and a viper's bugloss. The viper's bugloss caterpillar does not feed on the plant it was named after; this was a mistake made many years ago, but the name has stuck.

Then there are moths named after the time of year they fly. There is a November, a December and a March. There are May and July highflyers, but whether these moths really fly high I am not sure, but at least the time of year is correct. The August thorn flies in August, a month earlier than the September thorn, although both can be found together in some years. There is no January moth, but then there are hardly any moths about in January. The appearance of the herald and the spring usher mean that spring has arrived and the sight of an autumn green carpet indicates the onset of autumn.

What about the drinker; an alcoholic, perhaps? No, it is a moth whose big, furry caterpillars occasionally take up little drops of dew from the grass upon which they feed. This behaviour was first described in 1662, but sixty years later old Eleazar Albin, the first scientific, or at least the first careful, moth-watcher did not believe this, on the (reasonable) grounds that he had not seen it for himself. However, Albin did name another moth the lobster, after its caterpillar which he says, 'has some resemblance of a crustaceous fish'. Much earlier, in 1664, Thomas Moffet classed the lobster caterpillar with the devil's coach-horse beetle, naming it *vermis staphylinus*, adding that horses were killed if they accidentally ate one in the hay. But despite its beetle- or lobster-like appearance it is certainly a moth caterpillar, feeding on the leaves of beech and similar trees.

On summer days a small brown moth with yellow hindwings can often be found flying in meadows during the daytime. It is called the burnet companion because it flies at the same time of year and very often in the same places as the vivid black and scarlet burnet moths. Flying with it, also by day, is the black-and-white mother shipton, an extraordinary name for a moth. To see why it has such a name, look at the moth sideways on and focus on the forewing. With a little imagination you can see the wicked face of old Mother Shipton, that famous witch of bygone days.

The male oak eggar also flies by day, but the female is active at night although she spends much time sitting around on vegetation waiting for the arrival of a male. The word oak is commonly used in moth names, nearly always denoting the food-plant of the caterpillar. The eggar, however, feeds on bramble, hawthorn, blackthorn, broom and similar shrubs, and the only explanation of the name that I can think of is that the rich brownish coloration of the moth reminded someone of oak wood or perhaps dead oak leaves. But what about eggar? The name was first used in 1705 and was then spelt egger. Now an egger is an old word for someone who collects eggs, but what has this to do with the moth, which certainly does not collect eggs? Again we have to go back to Eleazar Albin. He remarks that the caterpillar fashions a silken cocoon, within which it turns into a chrysalis, in the shape of an egg, which it certainly does.

The goat moth, whose scientific name is *Cossus cossus*, is so called, the moth books say, because the caterpillar smells like a he-goat, but I have to admit that I could detect no such odour in the caterpillars I have kept. The caterpillars are big and fat and feed for three or four years on the solid wood of elm, ash, willow, poplar, birch, and other trees, and there are records of them feeding on timbers in buildings. P.B.M. Allan in his entertaining book *A Moth-hunter's gossip* tells how in 1724 caterpillars of the goat moth were found boring into the several-hundred-year-old oak timbers supporting the bells in Upminster church steeple. It is impossible to say what truth there is in this story, nor in another told by Allan, who warns:

'Beware of confining *cossus* even temporarily in a wooden box. There is a story of a certain entomologist who left a full-fed *cossus* for the night, in a wooden box, on the grand piano in the drawing-room. His wife, of course, was also an entomologist. And in the morning they heard the strains of Mendelssohn's "Spring Song" coming from the drawing room. Said the entomologist to his wife: "There's that damned goat moth." He was wrong, of course. It was his daughter, who had got up early to practise . . . but I can understand him believing any evil of a brute like *cossus*.'

There were (and perhaps still are) beliefs in certain parts of the country that some moths, or even moths in general, take the form of the human soul. One species in particular, called the ghost, has this reputation. The male is silvery-white above and dark below and when it flies at dusk it swings to and fro in a remarkable pendulum-like flight. Because of its coloration it seems to appear and disappear, hence the name ghost. The moth used to be called soul in parts of Yorkshire and people were scared of it and left it alone. It is also reputed to be especially common in cemeteries, but I can neither confirm nor deny this observation.

There was a time when many of the larger moths were called millers. These moths had another attribute in common besides their size: they were whitish and distinctly fluffy, rather like the species we now call the puss. One moth is still called the miller and is powdery white, especially around the head. Millers, the men, that is, often become dusted with flour, and were at one time believed to be dishonest, possibly because of the story of the miller in Chaucer's *The Reeve's Tale*. Long ago, if a large whitish moth was found, a Dorset child might recite:

> *Millery, millery, dusty poll,*
> *How many sacks has thee a-stole?*
> *Four and twenty and a peck*
> *Hang the old miller up by the neck.*

Then, when this had been chanted, the child would stamp on the moth and kill it.

The largest British moth, with a wingspan of 12 cm, is the death's-head hawk. It is a migrant from the continent, fairly frequent in some years, but rare in most, and is immediately recognisable by the Jolly Roger skull and crossbones marking on the back of the thorax. To the uninitiated it is a fearsome beast, with yellow and black hindwings giving it a superficial resemblance to a gigantic wasp. Added to this it can squeak, most unusual for a moth. It has the strange habit of entering beehives to feed on honey, and individuals are sometimes found dead when hives are cleared out.

The death's-head has always been regarded with awe, particularly in the nineteenth century. There is a record that one Sunday a moth landed on a pathway leading to a church. People going to church became scared, believing it to be a bad omen. Then someone decided that the best thing was to summon the village blacksmith to come and deal with it in an appropriate way. The blacksmith arrived, carrying his hammer, and taking a step back, gave it a great clout. The unfortunate moth was flattened on the pathway, which was considered a right and proper demise for such an unwelcome visitor. It is a pity, though, because the moth is completely harmless, except that just occasionally its large caterpillars defoliate potato plants.

Moths are named, or have been named, by consensus and by tradition. A name is coined and finds its place in the books and magazines. Sometimes it sticks, or a new one is suggested. There are no rules, as there are in the giving of scientific names, just a kind of gentleman's agreement. James Petiver in 1695 was the first to try and systematically name moths, and some of his names are still used, although many have changed, in my view not always for the better. For example, he called the moth we now call the cream-spot tiger, the yellow royal leopard, a much better name because the moth is not striped like a tiger but spotted like a leopard. Petiver ignored most of the smaller species which he collectively referred to as garden moths, and concentrated on the bigger, more spectacular species.

In 1771, and a little later in 1775, William Curtis and Moses Harris invented many of the moth names we now use. Even so some of Harris's names cannot be precisely traced to species. We do not know the true identities of his large goose egg, small old gentleman, and cross barred housewife, but what a sense of fun he and Curtis must have had as they coined names, and how well the tradition has been carried on. Thus, in 1870 an African moth new to Britain was captured at Tunbridge Wells. It was not recorded again until 1955 when one was captured in Surrey and another in Berkshire. It had to have an English name, so what to call it? Tunbridge Wells gem, of course, what else?

−9−
Tinker, tailor, soldier, sailor

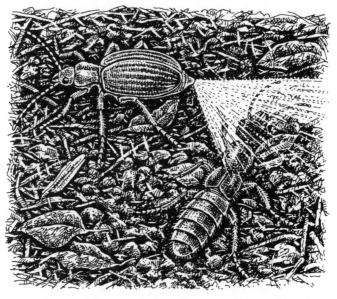

NATURALISTS HAVE BEEN quick to see or imagine they can see the resemblance between certain animals and man's professions and occupations. Hundreds of species bear common names which allude to the supposed similarity in appearance, voice or behaviour of an animal to a person going about his job or duty. There are tinker birds and tailor ants in Africa, and soldier flies and sailor beetles in the English countryside. Tinker birds are small barbets and the call of one species, the yellow-fronted tinker, is rendered in the books as, 'cok-cok-cok-cok-cok-cok . . .' given continuously and monotonously throughout the heat of the day, like a tinker of bygone days tapping metal pots and pans. It is so persistent that it is possible to be mildly puzzled by the sudden silence when one stops calling. Tailor ants cut leaves and sew them together with the precision of a skilled tailor cutting and sewing cloth, soldier flies sit in a stiff, upright position with the wings folded smartly down the back, while sailor beetles have an erect naval stance when at rest.

The clergy, in particular, have received due attention in the common names of animals. There are cardinals, bishops, friars, nuns and monks; also quakers (moths) and both conformists and non-conformists (moths again). The cardinal is a brilliant scarlet bird often seen at feeding tables in gardens throughout the eastern United States. The name has also been given to a bright red European beetle and an African woodpecker which, however, is disappointingly dull-coloured. A European shieldbug found in grassland is called the bishop's mitre, presumably because of its shape, while in Africa there is the large woolly-necked bishop stork.

Africa is particularly rich in other kinds of bishops, small birds, related to sparrows, so named because the males have a brilliant red or yellow and black plumage like the robes of a bishop, and perhaps also because they sit upright as if they are addressing a congregation or conference. They are sometimes called bishop-birds as if to avoid confusion with the real thing, but are more often referred to simply as bishops, which can lead to embarrassment when ornithologists write articles describing their behaviour. In 1935, Dr David Lack published an article entitled, 'Territory and polygamy in a bishop-bird, *Euplectes hordeacea hordeacea*' which sounds a little heavy-going. Rumour has it that when Lack submitted the article for publication he called it, 'Territory and polygamy in a bishop' and the editor of the magazine refused publication until the title was made more respectable. Recently, I saw an article, published in South Africa, with the title, 'Mate attraction and breeding success in the red bishop'. The political and religious overtones in this title sharpen when I look in my African bird books and find that besides the (communist?) red bishop, there is (understandably) a black bishop, a (Chinese?) yellow bishop, a (deviationist?) Zanzibar red bishop, and (to warn us of the perils of hell) a fire-fronted bishop.

Monarchs (a profession of sorts) occur throughout the warmer parts of the world. These large and colourful butterflies are most often seen in open country. One American species is called the queen (another profession) while the common species of the African savanna is called either the African monarch or the African queen. There are also monarchs in the rain forests of Africa which are black and white, not mainly orange like their open country relatives. One abundant forest species is called the friar and another the monk, and to complete the assembly there is also the novice and the layman; why yet another forest monarch is called the chief is not clear, unless it is an attempt to add a genuinely African dignitary to the group.

The pope, also called the ruffe, is a small, perch-like fish, common in English rivers. It might have been more appropriate to have named a much grander fish after this high church office. In New Zealand there is the parson-bird, a kind of honey-eater, decorated with two white feathers on either side of the throat. It has a melodious song, as befits a parson, but also mimics other birds. But pride of place in ecclesiastical names must go to the black-and-white apostle-bird of Australia which 'goes about in parties of twelve'.

The puffin, that dumpy little sea-bird with the improbable bill, has a curious scientific name in *Fratercula arctica*. The species name *arctica* denotes that it lives in cold northern waters, and is not especially imaginative, but the generic name *Fratercula* is from the Latin *fraterculus*, a friar, and is given because when a puffin flies up from the sea it keeps its legs together as if in prayer. Similarly, praying mantids, which are predatory insects related to grasshoppers, are named because they hold their big front legs together as if they are praying. A common southern European species is called *Mantis religiosa*.

The armed forces and the police are well represented in the common names of animals. Besides soldiers and sailors, there are adjutant storks, which stand erect like a junior officer, but it is a non-commissioned officer, the bombardier, that is more famous. The small bombardier beetle defends itself from attack by a predator, such as an ant, by squirting a hot, nasty-smelling spray from its anus. It does this with remarkable accuracy which makes it a very unpleasant insect indeed. There is also a rifleman, a New Zealand wren, a grenadier fish, and a kind of jellyfish called the by-the-wind sailor.

African butterflies of the genus *Precis* have been dubbed with a curious assortment of names which, it has to be admitted, have not yet gained wide acceptance as most people interested in African butterflies use scientific names when speaking or writing about them. *Precis* butterflies include the naval commander, the commodore, the brown commodore and, rather unexpectedly, the soldier commodore, which was common in my Sierra Leone garden. No doubt all of these would be interested in apprehending the pirate, a large, powerful-looking *Precis*, fond of drinking from fermented fruit, but the privilege might go to another species, the garden inspector, a butterfly that often enters gardens, but which I like to think of as a member of the local constabulary.

There are plenty of admiral butterflies: Ethiopian, short-tailed, long-tailed, orange, and, of course, our own red admiral, although it is often claimed that admiral is a corruption of admirable. Again, in Africa, there is the striped policeman, a kind of skipper (another profession), and all over the world there are copper butterflies, including the large, the small and the fiery, and, interestingly, the warrior copper of South Africa. I should at this point mention the pilot fish, named for a mariner, but why not stick to butter-flies?

Lurking in the side streets and lapping up nectar from attractive flowers is the painted lady. She could be visited by the red admiral or the grizzled skipper, thereby creating a dilemma for the large copper who may be prevented from investigating by the gate-keeper. Leaving geographical considerations aside, the large copper might be diverted by a Cambridge vagrant, an African butterfly, but he would have to take care because waiting around there could be assassin bugs, which have bearded faces and ambush their prey, or robber flies, which suck the blood of their victims.

There is a multiplicity of names denoting royalty of one kind or another. Besides a monarch and queen butterflies, there are kings (crows, eiders, vultures, penguins, flycatchers, snakes, fish and fishers), emperors (butterflies, moths, penguins and dragon-flies), a painted empress (an African butterfly), a Queen of Spain (a European butterfly, common in Spain), and Pharaoh's ant. Pharaoh's rat is another name for the Egyptian mongoose – highly regarded in ancient Egypt because it ate crocodile eggs – while Pharaoh's chicken is a name for the Egyptian vulture, a sacred bird in ancient Egypt, often figured in sculpture and esteemed as a useful scavenger.

The servants of royalty are commemorated in the footman moths, so named because they hold their wings down close to the body giving them an elongate and stiff appearance like a footman. There are a lot of species, including the dingy, feathered, hoary, pigmy and muslin, but my favourite is the red-necked – the one that might have dropped the soup? But what about the more mundane professions?

No profession is more mundane than that of a cleaner. About twenty-six species of fish, six species of shrimp and a crab obtain their food by cleaning and hence are called cleaners. Most species of cleaners are brightly coloured and conspicuous and advertise themselves by behaving in a rather flamboyant way. Their customers, larger fish, solicit the removal of parasites, growths and scabs by entering the territories of cleaners and allowing themselves to be cleaned. Cleaners may even enter the gill openings and mouths of their customers to remove food morsels.

The shoveller duck is so named because of its shovel-shaped bill which it uses to scoop food from the water. Carpenter bees make holes in wood, mason or potter bees make pots (hard cells) of clay and saliva and stock them with paralysed caterpillars to feed the larvae. Weaver-birds weave beautiful nests of grass and straw; many belong to the genus *Ploceus*, a Greek word meaning a weaver. The nurse shark retains its eggs in the oviduct and produces live young, unlike most sharks, while the larvae of the timberman beetle feed under the bark of fallen pine trees. Not to be outdone, there are tropical American woodpecker-like birds called wood-hewers, while in many parts of the world there are butcher-birds which kill and hang up their prey in larders, using the thorns and spikes on trees and bushes as hooks. Males of the European midwife toad entangle ropes of spawn around their thighs where it remains until the tadpoles hatch. The male midwife has obviously infiltrated a closely guarded female profession and rejoices in the name *Alytes obstetricans*. Music is represented by the trumpeter of South America, a bird which looks like a long-legged guinea fowl, named for its strange ventriloquial cries uttered with a closed beak. Pure science is not forgotten as there is a geometrician, a small English moth belonging to the family Geometridae, members of which are also called loopers and span-worms because of the way the caterpillars walk. Back-swimmers, also called water-boatmen, are boat-shaped bugs which swim by using their legs as paddles; forester moths may live at the edge of the forest but are more often found in the open, so the name is not an especially good one.

Better is the surgeon or doctor fish which has what amounts to a movable scalpel attached at the base of the tail. Predators are put off by the razor-sharp switch-blades which snap open when the fish is attacked. In Africa there is the secretary bird, a long-legged predator of snakes which looks like a tall chicken and is named because its crest feathers resemble the feathers of a quill pen used by a lawyer's clerk in the days when quill pens were still in use. The chimney sweeper is a sooty-black moth, fairly common in Britain, and the blacksmith plover of Africa is named because its call is a 'loud ringing, clinking cry, exactly like the sound of iron hitting stone'.

Angler fish have rod and line contraptions extending from the head to which 'bait' is attached. The 'bait' is actually part of the fish. Their enormous jaws swallow unsuspecting prey which are attracted to the bait as it moves gently in the water currents. The archer fish has a different technique. By compressing the gill covers it forces a jet of water through a tiny tube formed by its tongue and palate. The jet brings down prey from vegetation overhanging the surface of the water. The archer's accuracy is the more amazing since it aims from underwater and compensates for refraction, except when it sticks its snout out of the water.

Trails of driver ants scour through tropical forests regardless of what is ahead of them. They kill and devour virtually anything and are extremely ferocious predators. Step on a trail and they are up your trouser legs biting into your flesh with powerful jaws. They are also called army ants, but I like better the East African name, safari ants, as it seems just a little more sedate. Most individuals are workers which are females that contribute to the well-being of the colony but which never breed. There are also worker bees and wasps (females) and termites (males and females) which serve much the same function.

More dubious 'professions' include the alchymist, an English moth, the gipsy moth, introduced from

Europe to North America where its caterpillars defoliate trees, and the common joker, an African butterfly. Another African butterfly is called the pied piper because it is conspicuously black and white like a magpie, and yet another is the blue heart playboy. The bearded mountaineer is a hummingbird, and so is the fairy; there are also fairy shrimps. The Bohemian waxwing is a bird that moves around in an unpredictable manner as if it had no permanent home, while the old lady, a moth, stays at home and hides behind curtains in suburban houses. Hermit crabs also stay at home. They take over the empty shells of cockles, whelks, winkles and other molluscs and live inside, changing to a larger shell when they have grown too big. In North America there is a hermit thrush, which keeps to itself, and throughout the American tropics there are species of tiny, jewel-like iridescent hummingbirds called hermits which also tend to be solitary; among them are the bronzy, hairy, sooty-capped, hook-billed and little hermits.

Mandarins are ducks found not far from Whitehall. They were introduced from China, and are not to be confused with mandarin oranges. Sea-urchins are bristly little brutes and deserve their name. The stargazer fish sounds a little idyllic for a profession; so do damsels, delicate dragonflies, named because of their beauty, as is the demoiselle crane, with the extraordinary scientific name *Anthropoides virgo*, named not only for its elegant form but because of its high-pitched voice. Being a widow is certainly a time-honoured life-style. In the old days the poisonous black widow spider of America bit many a husband as he sat on the seat of the outside lavatory and sometimes may have left a widow. Also in America is the crazy widow, usually called the limpkin, a small, crane-like bird, named because of its melancholy call, and also known as the lamenting bird.

The scientific name of the Egyptian mongoose is *Herpestes ichneumon*. *Herpestes* is Greek and means a creeper, an allusion to the mongoose's stalking habits. The same word gives us herpes, that dreaded virus disease whose symptoms are creeping sores, and herpetology, the study of amphibians and reptiles, many of which are creepers. The specific name *ichneumon* is also Greek and means a tracker, another allusion to the mammal's stalking abilities. The word appears again in the Ichneumonidae, an enormous family of mostly small parasitic wasps which track down the larvae of other insects and lay eggs in their tissues.

The wryneck is a bird that can twist its neck into contortions, especially when it is alarmed, but it would be far-fetched to suggest it gives a wry smile or grimace. Its scientific name is *Jynx torquilla*. The specific name is derived from the Latin and means little twister, a slang expression for a marginal and suspect occupation, but then in Greece the wryneck was once used in witchcraft, hence the word jinx for a charm or spell.

−10−
Tools of the trade

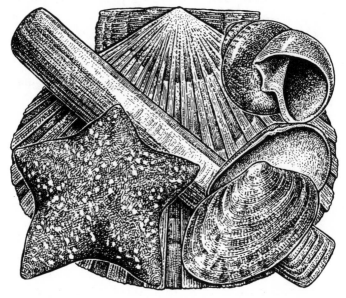

*'I am but mad north-north-west; when the wind is
southerly, I know a hawk from a handsaw.'*

AND I KNOW the difference between chalk and cheese
and can tell my left from my right hand. Such pro-
verbial phrases, often strengthened by alliteration, are
common in the English language and tend to be used
to repudiate a charge of stupidity or madness. But
what exactly did Shakespeare mean by the words
hawk and handsaw in these lines from *Hamlet*?

Mention of the winds suggests the outdoors, in which
case the hawk could have been a trained falcon and the
handsaw the heron − often hunted with falcons in

Elizabethan times. Handsaw or heronshaw, also written heronsew, harnser and in many other ways, might have been a hern's or heron's wood, a heronry, but is more likely derived from the Old French *heronceau*, a young heron. Everyone, especially outdoors people like falconers, would be expected to know the difference between hawk and heron, no matter what the direction of the wind.

But there is another possible explanation. A handsaw is a saw held in the hand by a carpenter, while a hawk is a wooden board held by the hawk-boy who stands beside the plasterer holding plaster ready for use. To not know the difference between such dissimilar tools would indeed be evidence of stupidity. The difficulty with this explanation is that there is no record of the use of the word hawk for a tool until 1700, but this does not of course prove that the word had not been used earlier.

No animal is named after hawk, the plasterer's tool, although there are of course many birds called hawks, but saw, the tool, has been incorporated into the names of several species. There is a saw-shark, a kind of dogfish with an elongate, blade-shaped snout armed with tooth-like structures. Sawfishes of the genus *Pristis* are also sharks found in warm seas in many parts of the world. They are unusual because unlike most sharks they occur in estuaries and also enter rivers and live in fresh water. The flattened, saw-like snout grubs up food from sand and mud at the bottom of the water, or is lashed from side to side in a shoal of fish in a clumsy attempt to disable one of them; it is also used in defence and is a formidable-looking weapon, although rarely directed against people. There are also sawbills, fish-eating ducks known as mergansers, which dive and hold the fish caught in the long saw-edged and hook-tipped bill. Two similar species, the red-breasted merganser and the goosander, are now common breeding birds in northern Britain, having greatly increased in distribution and abundance in the last thirty years. The sawflies are members of the order Hymenoptera (bees, wasps and ants) whose caterpillars cut into and chew the leaves of plants and whose females have sharp ovipositors for inserting eggs into vegetation. These caterpillars look very like those of butterflies and moths and not a bit like the larvae of bees, wasps and ants.

Many species of animal are named because of a supposed resemblance of part of their anatomy to a tool, implement, household article, musical instrument or weapon. In some instances the whole animal may be so named, and in a few it is the sound that is recalled in the name.

Related to saw-sharks and sawfish is the hammerhead, another shark which has a hammer-shaped head, as does the hammerkop, a heron-like bird found in Africa, believed by some tribes to be sacred and hence not molested. However, the yellow hammer is not named after a tool, the word hammer being derived from the Anglo-Saxon *amore*, a small bird, or the German *ammer*, a bunting, although why the bird is not simply called the yellow bunting is unclear, except that yellow hammer is a far more attractive name. An old Shropshire name for the bird is writing master, while from Northamptonshire there is scribbling lark, both names given because of the irregular lines on the eggshells which resemble writing or scribbling, a rare example of a bird named for the pattern of its eggs. John Clare (*Poems*, 1820) contributes this delightful verse,

> *Five eggs, pen scribbled o'er with ink their shells,*
> *Resembling writing scrolls, which Fancy reads*
> *As Nature's poesy and pastoral spells –*
> *They are the yellow hammer's, and she dwells*
> *Most poet like, 'mid brooks and flowery weeds.*

The spadefoot toads of Europe and North America live in burrows in dry places and dig with their hind feet which they use as spades. Unlike most toads they enter water only for a short period in the breeding season. The scientific name of the rabbit, *Oryctolagus cuniculus*, emphasises its digging habits as *oruktér* is Greek for a digging tool which, when combined with *lagós*, Greek for a hare, gives a digging hare. This is not quite an accurate description, because hares do not dig, but *cuniculus*, Latin for rabbit and for an underground passage, sets the record straight.

Filefish have compressed bodies, strong spines on the fins and tail and are not only very prickly but also poisonous. Several kinds of marine snail bore neat circular holes into oysters and other bivalve molluscs and suck out the body contents; they are appropriately called drills. Some bivalves are known as clams because the two-hinged valves clamp shut; the giant clam, which is supposed to be able to hold fast a man's leg, is over a metre in length. Wedge is a common prefix name, nearly always used to describe the shape of a bird's tail as in wedge-tailed eagle. Sickle also describes shape as in white-tipped sicklebill, a hummingbird, and the sicklebill, a kind of shrike found in Madagascar. The shoveller is a duck with a shovel-shaped bill. It was at one time known as shovel bill, broad bill, shovelard or spoonbill (but see below), also as whinyard, a name for a knife of the shape of a shovel.

Items from the kitchen are recalled in the knifefish and the forkbeard, a fish with long and filamentous pelvic fins, and in the spoonbill, a long-legged bird with a spoon-shaped bill. The many species of mainly tropical American oven-birds build oven-shaped nests of clay. There is a pepper-pot shell but the name is obviously rather fanciful as it is also called the watering-pot shell.

Turning to furnishings, there are dozens of species of carpet moths, including the beautiful, silver-ground and garden, all found in Britain, and named because of their pattern; there are also carpet shells and sharks, but carpet beetles are so named because they feed on carpets in a most destructive manner. The lamp-shells comprise a whole phylum, the Brachiopoda, and are named because the shell resembles a Roman lamp. There are about two hundred living and over two thousand fossil species. Brachiopods were among the earliest animals to appear on earth; fossils 500 million years old closely resemble living species, indicating a slow rate of evolutionary change in this phylum. Lanternfish live deep in the ocean and have luminous spots on their bodies; some species also have a large luminous spot on the tip of the snout, like a headlight. In contrast, there is the bed-bug which at night sucks blood from people as they lie in bed. It was at one time common in Britain but fortunately is now rather scarce.

The shoe-billed stork, also called the whale-headed stork, is a unique African bird and may or may not be a stork. It lives in papyrus swamps in Uganda and the southern Sudan and is rarely seen. The huge shoe-like bill is used for snapping up frogs and toads and it is clearly visible as the bird flies high overhead. The snowshoe rabbit is really a hare and lives in the sub-Arctic and Arctic of North America. It is also called the varying hare because its reddish-brown summer fur changes to white in winter, enabling it to match the snow and escape detection. The umbrella-bird is a black cotinga found in South America. It has a huge umbrella of a crest which projects beyond the bill. The umbrella-tree of Africa has umbrella-shaped leaves. It colonises wherever forest is cut down or where a big forest tree has toppled over. At Makerere University in Uganda, umbrella-trees were planted in a quadrangle and as they grew a butterfly called *Acraea pentapolis* colonised them. The butterfly caterpillars defoliated the trees, climbed up walls of buildings and entered lecture rooms. It is said that former President Idi Amin was so disgusted by the mess made by the caterpillars that he gave orders that the matter should receive instant attention which, needless to say, it did.

Continuing with domestic goods and chattels, there are cushion-stars, small starfish that sit about on rocks and look like diminutive cushions, and the hat-pin sea-urchin, *Diadema setosum*, covered in long spines that can penetrate the human skin, and which is common on the Great Barrier Reef off the Australian coast. The sand-burrowing razorshell, a bivalve mollusc, looks like an old-fashioned cut-throat razor which has been closed into its case; the shells are often found on sandy beaches but the animal is hardly ever seen alive as it lives deep in sand. Box-turtles and box-fish are shaped like little boxes, and basket stars, which are starfish, are similar to small baskets. There are also button quail, comb-oysters, necklace shells, needlefish and needle-whelks, and the mirror carp, a domesticated form of the common carp which has very large, mirror-like scales along its sides.

Then for travel there are the paddlefish of North American and Chinese rivers. They are up to two metres in length and have long, paddle-shaped snouts used to stir up the mud to find food. The tent-tortoise has a carapace thrown up into a series of conical bosses, looking for all the world like a group of camouflaged tents pitched by an advancing army. There is also an oarfish with the remarkable alternative name of king of the herrings, a canoe-shell and a boat-billed heron. The extraordinary petroleum flies, *Psilopa petrolei*, lay eggs in thick crude oil and the resulting maggots swim in the oil and feed on insects trapped in it. The oil-bird of northern South America is related to the nightjars and nests in colonies in caves, feeding its young on oily fruits gathered in the nearby forest. The carcasses of the nestlings were once impaled on sticks and used as torches by the local Indians as they travelled through the forest.

Musical instruments are well represented in animal names. The organ-pipe coral is shaped like the pipes of a church organ, as is the organ-pipe cactus, a common plant in the arid regions of Central America. The Australian lyrebird has a tail shaped like a lyre. It is also remarkable for its ability to imitate sounds made by mammals, including a human whistle, also the whistle of a train and the noise made by a chain-saw. The sound of music, not just the appearance of musical instruments, is recalled in the bell-magpies, Australian song-birds, of which one species, the western bell-magpie, calls in a chorus of bell chimes.

The white bell-bird of Guyana has a black, fleshy caruncle on the forehead. When excited it raises the caruncle and utters a sound like a bell. Rattlesnakes occur throughout the Americas and there are many different species. The rattle at the tip of the tail is made up of a series of loosely connected horny segments which when shaken make a sound like a hand-rattle. The sound frightens away potential predators and warns that the snakes are poisonous. Drum fish, also called grunts, occur in warm seas and make a sound with their air bladders that is said to remotely resemble drumming.

Finally, weapons of war, for which perhaps not astonishingly there is a particularly rich selection of animal names. The Portuguese man-o'-war is a big jellyfish, common in the Atlantic and occasionally seen in British waters. Its poisonous tentacles stretch for many metres and can inflict an unpleasant sting. Hidden among the tentacles lives a small fish, the Portuguese man-o'-war fish, which is unaffected by the poison. This fish is never found free-living; it is always in association with the jellyfish, and not only gains protection from its peculiar behaviour but also obtains fragments of food left over by the jellyfish. Man-o'-war birds, also called frigate birds, are found in tropical waters and sometimes stray north. They have wingspans of up to two metres but weigh only about 1.5 kilograms. This gives them unrivalled powers of effortless flight and enables them to swoop on and attack other sea-birds to force them to give up food; they are true pirates of the open ocean.

Especially well named is the sabre-toothed blenny, a small black-and-white fish that mimics in coloration cleaner fish of the genus *Labroides*. The cleaners' customer fish are tricked into receiving the deceptive blennies, but instead of being cleaned they end up with pieces of flesh ripped from their bodies by the ferocious blennies.

For attack, there are cutlass-fish with fierce teeth, and swordfish with extremely long upper jaws. There is also a sword-tailed hummingbird and there are sword-tails, popular aquarium fish related to guppies which have been selectively bred to produce in the males an enormous variety of elongate, sword-like anal fins. For defence, there are shieldbugs shaped like shields, helmet-shrikes with stiff feathers on the forehead which project forward over the nostrils, and the hooded visor-bearer, a hummingbird with a visor.

Arrow worms are, roughly speaking, arrow-shaped and constitute a large part of the oceanic plankton; they belong to a distinct phylum, the Chaetognatha. The bowfin is a primitive fish found in the Mississippi and the Great Lakes of North America and is named because of the long bow-like fin along its back. This is not an especially imaginative name and is hardly worthy of comment, but one of its common names is grindle or John A. Grindle, a rare example of the name of an animal incorporating the middle initial of a person's name.

Naming animals after tools, implements, household articles, musical instruments and weapons has obviously been a major preoccupation. It is certainly possible to see resemblances, some of them appropriate and accurate, others more fanciful or dubious. But we have been equally quick to borrow names from plants and animals for our own purposes, as I shall explain in the next chapter.

−11−
What shall we call it?

MY MOTHER'S MAIDEN name is Rose Nightingale; I have never come across anyone with such a striking natural history name, apart from her sisters Violet, Daisy and Lily. Naming a daughter after a flower is commonplace but is rarely combined with a bird surname, especially a beautiful songster like the nightingale. Besides the first names of my mother and my aunts I can think of Poppy, Primrose, May, Marigold, Heather, Pansy, Myrtle, Iris, Holly, Ivy, Bryony, Fleur and even Sorrel, which is not a good choice as the plant is so sour. Erica, which is also the generic name for heather, is apparently a diminutive of the masculine Eric and not a scientific name used for a girl's first name, as I had first thought. Rosemary is a combination of Rose and Mary and not from rosemary the flower, which is so named because it is often a maritime species. Prunella is a fairly common girl's name; it also happens to be the generic name for both the dunnock, a bird, and self-heal, a plant. The name could mean a little prune, hardly appropriate for a girl's name; it is also an altered form of brunella, an old name for an infectious disease which, some believed, should be treated with *Prunella*, the plant. Prinia and Dafila are unusual, possibly unique, girl's names, given by two well-known ornithologists for their daughters. *Prinia* is the generic name for a group of African warblers, while *Dafila* is the generic name for the pintail duck. It should not be difficult to guess the identity of the two ornithologists.

Countless names of animals and plants have been and still are borrowed and used for inanimate, man-made objects as diverse as cars and weapons of war. And we not only name our daughters after flowers and birds but persist in seeing similarities in human behaviour and the behaviour of animals in general or in particular. Carelessness in dubbing people and things with animal names is widespread, pervasive and for me occasionally annoying. Only today I read in the newspaper (yet again) of drunken football fans behaving like animals. Derogatory comparisons abound: silly cow, dirty dog, stupid ass, cheeky monkey and guttersnipe; others offer a more accurate and reasonable comparison as in cunning old fox; yet others are purely descriptive as in bald as a coot (coots have bald pates).

What shall we call it? Well, in slang at least the answer is virtually anything. Prattling, saucy talkers, mostly female, may be dubbed chatterjays or chatter-pies; magging is a slang word for chattering, and to chatter noisily in a mildly scolding manner is to mag. All these words are from magpie, a noisy, chattering bird, familiar to everyone. Magpie is also a slang word for an Anglican bishop dressed in black-and-white vestments; a piebald eye is a black eye, itself a slang expression.

To refer to a lady as a spinster would these days be regarded as rather old-fashioned; feminists would surely condemn the word as sexist and chauvinistic. The word was first used descriptively to mean a woman (rarely a man) who spins, but since the seventeenth century has been used as the legal definition for an unmarried woman. Subsequently, and perhaps rather unkindly, it came to mean an old maid or someone who behaves like an old maid. Spinster has the same origin as spider, both words meaning spinner. The Old English *spinnan*, to spin, gives spinster, also *spinthra* or *spithra*, thence to Middle English *spithre* or *spither*, subsequently altered to spider. According to S. Potter and L. Sargent in *Pedigree: words from nature*, *spithre* and similar words were eased to spider because of difficulties of pronunciation. The change makes pronunciation less difficult as, for example, in the alteration of murther to murder.

Modern English usage abounds with the names of things borrowed from the names of animals: there must be as many tools and weapons named after animals as there are animals named after tools and weapons. A monkey-wrench is a kind of spanner; a crowbar is an iron bar with a wedge-shaped end, usually slightly bent like a crow's beak, and used as a lever; at one time it was simply called a crow. Tigers, panthers, leopards and elephants are armoured tanks, named to emphasise strength, speed or size; armoured cars and combat vehicles include the puma, fox, ferret, vixen and scorpion, presumably named to indicate stealth, cunning and stinging fire-power.

I can think of over forty kinds of aircraft named after animals; there are probably many more. There are puma, gazelle, wasp and lynx helicopters, perhaps named for their light-footed ability to take off and land on a small space. It might be expected that many aircraft would be named after birds, especially the more graceful flyers, and there is certainly a good range of bird names, including skua, heron, eagle, kestrel, albatross and condor. The harrier is a fighter, capable of vertical take-off and landing and able to remain still in the air; it can even fly backwards. It is obviously and aptly named after harrier, a name given to several similar species of birds of prey which can also take off and land vertically and which are capable of quite remarkable aerial manoeuvres, especially when hunting prey, but also during courtship. The word harrier really means a catcher of hares; the marsh and hen harriers occasionally kill hares, especially young ones, but their main prey consists of smaller animals such as rodents and aquatic birds which they pounce upon and kill. The word harrier is also used for a kind of hound resembling a small foxhound and used for hunting hares; a person who hunts with these hounds is also called a harrier, as might a sprinter or runner who belongs to a club like the Blackheath Harriers.

But back to aircraft, for which human inventiveness has produced a rich variety of names borrowed from animals. There is (I should probably write was, as I am not sure all can still fly) a mustang, walrus (a seaplane), sea-vixen (a naval jet), vampire, jaguar (also a name for a car), hell cat, chipmunk, beaver, bear cat, buffalo, camel and bulldog, all mammals, some of the names most improbable, at least for me, but then I am not an aircraft buff. I hear also that NATO code names for Soviet military aircraft include bison (another name for buffalo), bear and foxbat. The tiger moth is a famous light aircraft which has earned for itself a significant place in the history of flying; so has the mosquito, a Second-World-War fighter that certainly had a sting in its tail. The Red Arrows are the Royal Air Force's aerobatic team of small, red jets whose pilots perform breath-taking twists and turns for thousands of admiring spectators. They used to perform their aerobatics with gnats but have now changed to hawks. I hear there is an aircraft called the flying flea, but cannot confirm whether this is really so.

Many cars, trucks and lorries have been named after animals, mostly after mammals, including the so-called 'panda' cars of the British police, and the greyhound bus service which provides fast transportation across the length and breadth of the United States. Names given to weapons of war are intended to highlight efficiency and destructiveness, and to put fear into the hearts and minds of the enemy. Not all have such names: the limpet mine can be fixed to a sloping structure, just as the limpet clings to wave-battered rocks. The torpedo was originally a box loaded with gunpowder which was placed in the water near enemy ships; only later did it become the sophisticated, streamlined projectile which could be fired at distant ships with formidable accuracy. The torpedo ray, from which the weapon is named, can produce an electric discharge which numbs its prey. It is therefore interesting that torpedo has come to mean a person who has a numbing influence on others. To torpedo is to paralyse, numb or destroy: politicians sometimes torpedo each other's policies.

Modern missiles are eminently suitable to be named after animals. Sea-wolf sounds just right for a missile; so does sidewinder, named after the snake that winds sideways as it progresses across the dry desert sand. Of recent notoriety is the French-made Exocet which is fired from ships or aircraft and which skims over the sea surface until it strikes the target with devastating effects. It is well named from flying fish of the genus *Exocoetus* which belong to the family Exocoetidae. These fish have enlarged pectoral fins which are used for gliding, not flying in the strict sense; interestingly, they glide to escape, not to attack.

The *Oxford English Dictionary* defines a mole as:

'A small animal about six inches in length, having a velvety fur, usually blackish, exceedingly small but not blind eyes, and very short fossorial forelimbs with which to burrow in the earth in search of earthworms and to excavate the galleries and chambers in which it dwells.'

I like the gentleness, indeed the touch of pathos, in this definition: the mole sounds so innocent and harmless. The *Dictionary* admits to a transferred meaning for someone who works in darkness, but there is no mention of the word to mean what amounts to a spy working to dishonestly obtain secret information for an enemy or competitor.

-12-
Camberwell remembered

A SIMPLE AND convenient way of inventing a scientific name for a species is to commemorate the place it was first found or the area of its geographical distribution. Many specific names end with *-ensis*, as in *Branta canadensis*, the Canada goose, or in *ana*, as in *Periplaneta americana*, a cockroach which, it so happens, is probably African in origin, not American, and is today found all over the world. These Latin suffixes simply mean 'coming from'.

Common names that denote a place may be taken directly from scientific names, as in the Canada goose; alternatively they may be coined when a species suddenly turns up in a place where it has never been found before. The Camberwell beauty is in Britain a rare migrant butterfly from Scandinavia and was first collected in August 1748 at Cool Arbour Lane, Camberwell, now a crowded suburb in south-east London. The butterfly is also called white petticoat because of the white or yellow borders to the wings which look like a petticoat peeping suggestively below the hem-line of a dress; in North America it is known as the mourning cloak because of its dark coloration. But at least in England, Camberwell is remembered in the common name of a magnificent creature which every butterfly enthusiast would dearly love to see.

Four other British butterflies have well-established common names that commemorate places. The Lulworth skipper is in Britain effectively confined to the Dorset coast where it is particularly abundant around Lulworth Cove. In some years it swarms on grassy slopes in August but is rarely noticed by the thousands of holidaymakers to this delightful spot because it is so small and ordinary-looking. The Essex skipper is likewise inconspicuous and is named because it is common on the marshes of the Thames estuary, but is also found at many inland localities to which it appears to have spread in the last hundred years or so. The butterfly is certainly an opportunist; some years ago it was introduced into North America and is now spreading rapidly across the continent. The Bath white, a common European butterfly but a rare immigrant to Britain, is apparently named because a young lady living at Bath executed a piece of embroidery incorporating the pattern of the butterfly's wings. Whether the Bath white has ever been recorded at Bath I am not sure, but the name has stuck despite the rather curious circumstances of its origin. Then there is the Scotch argus, also found at a few English locations, but virtually confined to Scotland where in places it is exceedingly common.

Several other British butterflies have in the past been given names after places. Perhaps the most dubious is the Arran brown which may or may not have genuinely occurred on the Isle of Arran. Entomologists still argue as to whether this species should be regarded as British, some even believe there could be undiscovered colonies of it in the remoter parts of Scotland. Albin's Hampstead eye was caught in the eighteenth century on Hampstead Heath by Eleazar Albin and illustrated by James Petiver. We now know (or think we know) that the butterfly is *Precis villida* from the Far East or Australia and was probably accidentally introduced as a caterpillar or chrysalis in a consignment of fruit.

Earlier this century, F.W. Frohawk, a most careful observer of butterflies and an excellent artist, noticed that the shape of the egg of the white-letter hairstreak butterfly is unique. He therefore suggested the species should be transferred from its genus and placed in a new one that Frohawk wanted to call *Chattendenia* to honour Chattenden Wood in Kent where, in the 1950s, as I can remember, the butterfly was particularly common. The name was never officially accepted, which is just as well as the butterfly has probably disappeared from the area by now. The elms, on the leaves of which the caterpillars feed, have nearly all been killed by Dutch elm disease.

In Britain there are many more species of moths than butterflies and there is hence more scope for inventing names after places for them. There is the Paignton snout, the Lewes wave and the Brighton wainscot, all named after three well-known south coast holiday resorts; the Edinburgh pug, the Kew arches, the Ealing glory (Moses Harris's name for the species now known as the green-brindled crescent) and, with a musical note about it, the Manchester treble-bar. The counties are not forgotten: Sussex and Essex each have an emerald, the Essex species is so rare that it is completely protected by law; Kent has a black arches and there is also the Kentish glory which used to be called the glory of Kent, but is now extinct in the county; there is a Devonshire wainscot and a Devon carpet; a Cumberland gem and a Cumbrian umber. Rannoch, that famous moth-collecting locality in the Scottish Highlands, is remembered for its brindled beauty, looper and sprawler; Portland in Dorset for its riband wave and also for what is un-pretentiously called the Portland moth. The New Forest has its burnet; the island of Jersey its tiger; while on the Isle of Wight there is the Isle of Wight wave, and on Arran, the island with the dubious butterfly, is the Arran carpet. Scotland as a whole has the slender Scotch burnet and the Scotch annulet; Wales the Welsh clearwing; but England has nothing – perhaps she should claim the cosmopolitan as her very own moth?

Many of the place names attached to Britain's wild flowers and trees allude to introductions from abroad. Just over two hundred years ago the seeds of a plant called *Conyza canadensis*, now known as Canadian fleabane, were accidentally brought into London docks in a consignment of goods from North America. Some of the seeds germinated on nearby wasteland and since then the plant has spread over much of England, particularly along railways which in the nineteenth century were radiating and proliferating from all major towns and cities. The habitat created by the gravel needed to support railway lines was excellent for this plant. Recently it has spread along motorway verges for much the same reason. Its small, light seeds are equipped with a parachute of hairs and are hence easily drawn along in air currents created by speeding trains and road traffic; this has undoubtedly aided colonisation. The name fleabane comes from a related native species, *Pulicaria dysenterica*, a plant which was once used for destroying fleas (the generic name is from *pulex*, Latin for a flea), and also as a remedy for dysentery, a use reflected in the specific name.

The plant now known as Oxford ragwort escaped from the Oxford Botanic Garden where it had been cultivated as an attractive ornamental and, like Canadian fleabane, spread along railways during the nineteenth century. In the 1940s and 1950s this south European species colonised the rubble resulting from wartime bombing and it is now found growing on old walls, building sites, railway embankments, roadside verges – indeed just about everywhere, often flowering in places where few other plants survive. It blossoms from May until midwinter and is an attractive alien but, interestingly, rarely colonises pasture fields where its indigenous relative, common ragwort, can be a noxious weed.

The London plane, that huge tree with the peeling bark and sycamore-shaped leaves, often seen in squares and along roadways in cities, is a hybrid between two non-native species, the oriental plane of Asia and the western plane of North America. Some of the original hybrid individuals are still growing and are already the tallest deciduous trees in Europe.

Among native plants that commemorate place names are the Snowdon lily (with the decidedly Welsh-sounding generic name of *Lloydia*), the star of Bethlehem, named because of its star-like white flowers, and the grass-of-Parnassus, *Parnassia palustris*, not a grass but an attractive plant with veined, white petals, described in the first century AD by the Greek herbalist Dioscorides as an inhabitant of Mount Parnassus. Yorkshire fog sounds an improbable name for a grass. The word fog is dialect for a coarse winter grass that grows after hay is cut, although it could conceivably be from the Old Norse *fogg*, a limp and damp grass. It is not especially associated with Yorkshire but certainly grows well there on hillside sheep pastures.

There is a seaside umbellifer, common at Dover, with the strange name of alexanders, not named directly from a place, but from Alexander the Great of Macedonia. In the seventeenth century it was used to cure stomach trouble and was known as Macedonian parsley. Dame's violet, a member of the cabbage family, is not named for a dame but is from the French *violette de Damas* which was misinterpreted as *violette des dames*. The plant should probably be called damask violet as the name is from Damascus, the Syrian capital. In contrast, herb paris, *Paris quadrifolia*, has nothing to do with Paris, the city. The common and generic names mean herb of a pair, an engaged couple, a reference to the four leaves being set together on the stalk like a true-lover's knot – the emblem of betrothal.

A swede is a kind of turnip introduced from Sweden. Some years ago I was driving through the

Oxfordshire countryside with two Swedish friends when they saw to their astonishment a roadside notice announcing 'swedes for sale'. It took a few minutes to explain and to reassure them that England is not that barbaric. Spinach is probably similarly named as the word seems to come from *hispanicum*, something from Spain. The -barb part of rhubarb is from *barbarum*, Latin for foreign. All foreigners were barbarians because they were barbed or bearded. Our rhubarb probably comes from the Soviet Union where many people cultivate beards. Deadly nightshade, whose poisonous berries are edible to a few birds such as pheasants, was once known as devil's rhubarb, a severe warning of its toxicity.

The English names of European birds incorporate a curious mixture of places, some of which seem to be only marginally associated with the bird in question. The Egyptian vulture occurs over much of southern Europe and in many parts of Africa and is not especially Egyptian. The Sandwich tern, *Sterna sandvicensis*, still occurs near Sandwich in Kent but is widespread and common off most European coasts; in winter many go to West Africa. The Kentish plover is probably extinct as a breeding bird in Kent, but is an abundant Eurasian species, and the Dartford warbler has long forsaken Dartford, although it is still found in Dorset. Ural owls and Siberian tits, Siberian jays and Siberian thrushes have a ring of the boreal forests in their names, while the Lapland bunting, or Lapland longspur, as the Americans call the bird, conjures up images of the high Arctic tundra. More precisely named is the Corsican nuthatch, confined to Corsica, but rather carelessly named is the Manx shearwater. This bird can be found at sea more or less anywhere west of the British Isles, north to Iceland and south to the Bay of Biscay; the French call it *puffin des Anglais*.

The Americans have made a point of naming birds after their states and indeed each state has a bird as an emblem, not necessarily the bird bearing the state's name. Connecticut, Kentucky and Tennessee each have warblers named after them; Florida has a jay and a gallinule, which is the same species as our moorhen; Louisiana a heron and a water-thrush; Mississippi a kite; California a gull; Oregon a junco; and Virginia a rail. There is also a Carolina chickadee and a Carolina wren, but whether North or South Carolina is not specified. Interestingly, a number of American birds are named after towns and cities in much the same way as British moths have been named, but probably not for the same reason. There is the Baltimore oriole, Ipswich sparrow, Nashville warbler, Philadelphia vireo and the Savannah sparrow. I like to think this bird commemorates the attractive seaport of Savannah in Georgia but I suspect it is named because it lives in open, grassy country.

In 1981, the United States Postal Service issued a delightful set of fifty 20-cent postage stamps, one for each state, each depicting the state bird. What is extraordinary is that of the hundreds of species of birds resident in the United States, only twenty-six species are represented. Moreover, two of these are chickens, the Rhode Island Red for Rhode Island and the 'blue hen chicken' for Delaware, while another, for North Dakota, is the pheasant, an introduced but much appreciated species. Seven states, all in the east, select the cardinal, a woodland species familiar at garden bird-tables in winter. Six states, all in the west, have the western meadowlark, a conspicuous bird of the open prairie, but the eastern meadowlark, an almost identical species, does not feature among the birds selected by the eastern states. The mocking-bird is claimed by five southern states; other species represented more than once are the (American) gold-finch (three states), the (American) robin (three), the bluebird (two), the mountain bluebird (two) and the chickadee (two). Not counting the pheasant and the two chickens, seventeen species feature once only, including some good choices like the Carolina wren for South Carolina and the Baltimore oriole for Maryland. The California gull is hailed as Utah's state bird, which at first seems a strange choice, but in the nineteenth century flocks of these birds are reputed to have destroyed hordes of crop-eating grasshoppers and so saved the people from starvation.

Non-native species, especially if they become pests or behave like uninvited aliens, are often dubbed with the name of their place of origin. In Britain we sometimes find Colorado beetles, serious pests of potatoes, and we are compelled by law to report their presence. Many would agree that we have far too many American grey squirrels, a species first released in 1890 at Woburn Park, Bedfordshire, reputedly because the local gentry had run out of things to shoot. The 'American' part of the name is nowadays dropped as there is no possibility of confusion, there being only one other squirrel, the red, resident in Britain.

Vagrants from across the Atlantic that arrive of their own accord tend to receive the same treatment. The American painted lady, a separate species from 'our' (actually worldwide) painted lady, is occasionally blown across on westerly gales; so is the American robin, a kind of blackbird not a bit like the robin of

Europe, but which the Americans simply call robin as the European robin never makes it across the Atlantic. Certain European birds, such as the black-headed gull, often find their own way across the Atlantic, but it is the deliberate introductions of plants and animals to North America that are the most impressive. Hundreds of species have been introduced, most of them plants and insects, some of which have become important weeds and pests. One such pest is the starling, formerly known by the Americans as the European starling but now generally known simply as the starling. Likewise the introduced house sparrow, dubbed by many as the English sparrow, has now reverted to its original common name; but the introduced goldfinch is still known as the European or British goldfinch to distinguish it from the native goldfinch which, it so happens, is a very different-looking bird.

There is only one species of wren in Britain and because there is no possibility of confusion the bird is simply called the wren. In North America the same species is known as the winter wren, partly because over much of the area it is a winter visitor from the north, but mainly to differentiate it from the many other species of wren found in North America, some of them, like the Carolina wren, named after places. This is an interesting example of the lengthening of a common name where there are several related and similar-looking species present in the same area, a practice that crops up again and again.

Some names from places have really strange meanings and origins. The Komodo dragon is not a dragon but it does happen to be the largest species of monitor lizard known. It was discovered in 1912 on several small islands, including Komodo Island, in the Sunda archipelago. The lizard may reach more than three metres in length and feeds on young deer, pigs and, it is alleged, on horses. An extinct relative, known only from fossil remains in Australia, was up to seven metres long, and must have been a really impressive lizard. The slight alliteration resulting from combining Komodo with dragon produces a suitably awe-inspiring name for this remarkable creature.

More circuitous is the name canary for the commonly kept cage-bird. Canaries are descended from greenish finches found on the Canary Islands. The birds have been selectively bred by aviculturists and now occur in an immense variety of plumage types, many of which are a pale yellow called canary-yellow or canary. The Canaries are so named because of the feral dogs that used to live on them, the dogs having escaped from visiting ships. The name is from *canis*, Latin for dog, so the Canaries could be called the Isles of Dogs. In this instance the name for both the bird and the colour come from the Latin word for dog.

–13–
Magpie names

My friends know that I dislike beach holidays. Sitting on a beach in the full glare of the sun with nothing constructive to do is for me a most unattractive prospect. The trouble is that beaches seem relatively devoid of wildlife; true, in winter when there are no people around, there are waders and gulls to look out for, but in summer when holidaymakers are out in force, all the interesting birds leave before breakfast and return only when it is dark and peaceful once again. If I happen to get caught up in a beach party I usually try and sneak off to look for sea shells washed up at the tide-line, an interesting but not entirely satisfactory way of passing the time because many of the shells are severely worn or broken by wave action and lack the lustre of live individuals. Moreover, I am never quite sure where they have come from as they are deposited on the beach by the incoming tide, and like all zoologists I really do want to know exactly where an animal lives.

One of the commonest shells found washed up on sandy beaches is the bivalve, *Donax*, known by a number of common names, among them the butterfly clam because the two symmetrical, hinged valves give the shell the appearance of a butterfly. In live butterfly clams the two valves fit closely together and separate to give the characteristic butterfly shape only after the animal's death. And then after a further few days the hinge breaks, the two valves separate, and the resemblance to a butterfly disappears. But for a short time, at least, the name is a good one, especially as the valves are often brightly coloured.

Living butterfly clams of various species (usually just one species to a beach) are found on sandy shores in most parts of the world, nearly always at or just below the low-water mark, although in relatively tide-free seas, such as the Mediterranean, they occur in shallow water. The animals remain buried in the sand and are hidden from view except when uncovered by wave action, and then they burrow rapidly back into the sand. They feed by extracting small organic particles and in some places are extremely common, an indication that despite the barren appearance of most beaches there is abundant life, albeit micro-scopic, at or just below the low-tide line. The best way to find living specimens is to stand near the low-tide line and make holes in the sand by 'puddling'. Butter-fly clams will then be washed into the holes.

My first encounter with butterfly clams was in 1957 in the extreme south-west of Portugal. I was in a small restaurant and ordered some mussels which turned out to be butterfly clams cooked in butter and flavoured with garlic, a delicious late-evening snack. Since then I have collected butterfly clams from beaches as far apart as Tanzania, Sierra Leone, the eastern United States, Greece, Spain and England, but it was not until 1984, again in south-west Portugal, that I tried ordering mussels in a restaurant and to my delight received butterfly clams cooked in butter, this time flavoured with parsley.

Some of the species of butterfly clam, notably from East Africa and the coasts of North America and the West Indies, are so variable in coloration it is difficult to find two individuals alike. This variation, involving beautiful combinations of white, purple, blue, orange and brown, with or without distinctive ray-like markings, makes them immensely attractive and for me compulsive collecting – I must have thousands of shells and every time I am persuaded on to a beach I collect more, a task that makes beach-life bearable.

Many animals are named after other, totally unrelated species, usually because of a fanciful resemblance, as with the butterfly clam, or a supposed similarity in behaviour, stance or voice. I like especially the barnacle goose, because there is also a goose barnacle. Then there is a crab spider and a spider crab, a mouse deer and a deer mouse, but such reciprocal names are rare – there is no clam butterfly.

Interestingly, there has been considerable carelessness in naming animals after other animals. For example, the word 'fish' is widely used for animals that are not fish. Thus shellfish (which, incidentally, may or may not have shells) are edible molluscs like oysters and whelks, also cuttlefish, and crustaceans such as shrimps, crabs, lobsters and crayfish. All these have in common with real fish is that they live in water and are edible. Silverfish, on the other hand, are not edible; they are insects, active at night, and often seen scuttling away on a cellar floor when a light is suddenly switched on. Apart from being silvery and rather long they bear not the slightest resemblance to fish.

The word 'worm' should strictly be applied to the phylum Annelida which includes earthworms and their marine counterparts, the polychaetes. I can, however, think of dozens of other kinds of animals commonly called worms which have no relationship to annelids. Some are certainly long and thin like an earthworm, but the word worm is also used in a derogatory way to indicate unpleasantness or something causing harm, as in tape-, round-, pin-, thread-, hook- and lung-worms, all parasites of man and other mammals and causing disease and suffering. Glowworms, wood-worms and meal-worms are beetles, while ship-worms are bivalve molluscs that bore into wood and damage boats.

Many moth caterpillars are called worms. Span- and inch-worms are caterpillars of geometrid moths, named because they walk by spanning their own length with each movement, moving about an inch at a time. The tobacco horn-worm, the caterpillar of a large hawk-moth, has a horn at its rear; in the United States it is sometimes a serious pest of tobacco plants. Web-worms live in silk webs spun on bushes, bagworms live inside bags of twigs spun together with silk, army-worms defoliate vegetation before marching off to a new patch, while cut-worms cut off plant stems at the base causing the plants to wilt and die. The silk-worm is a domesticated caterpillar which when fully fed spins a silken cocoon from which commercial silk is spun.

A few vertebrates are called worms. There is the worm-lizard and the slow-worm, which looks like a snake but is really a legless lizard. To call someone a worm is very rude indeed as it implies contempt and insignificance, unless of course the person is called a bookworm, which can be taken as mildly compli-mentary.

If an animal is boldly and distinctly patterned its name is immediately liable to be used for a host of similarly coloured species. The black and white of the magpie, together with its familiar appearance in the European countryside, has resulted in its name being borrowed for many other species including magpie goose, magpie robin and magpie lark, all of them birds of distant lands and named by British travellers who saw something of the magpie in their coloration. In England there is a magpie moth, also called the currant moth because its caterpillars feed on the leaves of currant bushes, and its much scarcer but similar relative, the clouded magpie. Unrelated to these two is the small magpie moth, in Britain abundant wherever there are stinging nettles, which means virtually everywhere.

When in winter plumage the red-throated diver is black and white, and in Norfolk, where the bird is only seen in winter, it used to be known as magloon. In Staffordshire and Hampshire the great-spotted wood-pecker, another black-and-white bird, was sometimes known as wood pie, while in Leicestershire it was called the French pie, French again meaning strange or alien; the similar but smaller lesser-spotted woodpecker was known as the little wood pie in parts of Hampshire. In Ireland, and also in Kent, the smew, a black-and-white diving duck, was known as the magpie diver. Yet another black-and-white bird, the great grey shrike, was called the murdering pie because it impales its prey (large insects and small mammals) on thorns; in East Anglia the lapwing used to be called horn pie, horn referring to its conspicuous crest; sea-pie (sometimes corrupted to sea piet, sea pilot or pienet) is still used locally for the oystercatcher; and an old Scottish name for the guillemot is maggie.

Magpie is an old name for a familiar bird almost universally disliked because of its reputation for stealing eggs and nestlings of other birds, including gamebirds, and because it was at one time considered a bird of evil omen. The name magpie may have been derived from magot or margot pie and it is, or has been, known by a large variety of folk and provincial names including madge, margaret, miggy, nanpie, ninut, pye mag, pie nanny, chatterpie, hagister or just plain pie. The names may all stem from Margaret (or some other version of this name), a chattering lady. Some authorities think the word pie for a dish of food is derived from magpie, the connection being that just as the magpie is supposed to collect assorted oddments for its nest, the cook assembles an assortment of ingredients for the pie.

Another derived word is pied, with means black and white like a magpie. Pied is used to describe dozens of black-and-white mammals and birds including the pied wagtail and pied flycatcher; also piebald, meaning much the same and most often used for a black-and-white horse, but also for a mongrel, or even a person of uncertain origin because of his or her mixed or hybrid appearance.

Magpies are usually seen singly or in twos, sometimes up to eight together, but in Britain never in large flocks as they are resident birds and do not migrate. The actual number of magpies seen together is in some mysterious way said to predict a coming event, as in the following North Country verse:

> *One is sorrow, two is mirth,*
> *Three a wedding, four a birth,*
> *Five in Heaven, six in Hell,*
> *Seven the deil's ain sell.*

The last couplet may be rendered:

> *Five a sickening, six a christening,*
> *Seven a dance, eight a lady going to France.*

Or in Lancashire as:

> *Five for rich, six for poor,*
> *Seven for a witch, I can tell you no more!*

There are various other versions, all along the same lines. Clearly, the magpie is an important bird and small wonder its name has been borrowed and used for so many other species.

The mammalian equivalent to the magpie is perhaps the zebra. There are zebra fish, finches, mice, spiders and swallowtail butterflies, and one South American nymphalid butterfly is simply called the zebra. This species has the distinction of being on the British list of butterflies because one emerged from a bunch of bananas at Eastbourne in December 1933, clearly an import.

The word harlequin is Italian in origin and in a sense can be viewed as an animal name because it refers to a character in comedy, a kind of clown dressed in parti-coloured, bespangled tights, usually carrying a wand. The name has been given to a variety of extravagantly coloured animals such as the harlequin duck, quail, lizard, snake, deer, bat and beetle. It is also used for the Dutch mastiff dog, and in England a local name for the magpie moth is the harlequin moth.

Apart from insects it is fish that seem to have received the most names borrowed from other animals. Butterfly fish are small and brightly coloured, with deep, flattened bodies; they are found in tropical waters in many parts of the world. A mirror image would give an immediate impression of a butterfly, but to get this effect it would be necessary to split one down the middle and lay the two halves together, much like a freshly dead and opened butterfly clam. Frog fish are also found in tropical seas, usually where there are sandbanks or coral reefs; they drift around among seaweed and are almost impossible to see because of their coloration and shape which someone must have likened to that of a frog. A quite extraordinary number of fish are named after mammals. Dogfish are probably so called because they are deemed common or vulgar and of no particular value, as in the plants dog violet and dog rose. Catfish have long, sensitive barbels like cat's whiskers, lion fish have long and poisonous spines, while porcupine fish inflate their air sacs and blow themselves up into a spiny ball whenever they are threatened by a predator. Other borrowed names for fish include goat, horse (a kind of mackerel), rabbit, squirrel, whale, rat and wolf. Among the sharks and their relatives is a tiger, a cow and a cat, and there is also a leopard and an eagle ray.

Interestingly, but perhaps not astonishingly, mammals have not been named after fish; indeed rather few mammals have names borrowed from other animals. The strange, egg-laying duck-billed platypus of Australia has a decidedly 'duck bill'. This mammal is also called a duck mole or water mole because it burrows and has short velvety fur like a mole. Bulldog bats are hare-lipped, long-legged, fish-eating bats, and mastiff bats are snub-nosed and described as being thick-set, like a mastiff dog. Spider monkeys of South American rain forests have exceptionally long legs and tails and move about the canopy like gigantic spiders.

Most borrowed bird names are from other birds, and there are many of them, almost as if ornithologists ran out of ideas for names; most refer to superficial resemblances. Thus the African cuckoo falcon is coloured and patterned like the European cuckoo which winters in Africa, but the falcon is quite unlike most of Africa's resident species of cuckoos. Its scientific name is *Aviceda cuculoides*, the specific name meaning cuckoo-like. Also in Africa there is the painted snipe, not a true snipe, and unusual because the female is the more brightly coloured. My bird book says it makes a 'hissing, swearing noise during display, also a guttural croak' and adds that the call of the male is shriller. Other misnomers include the European hedge sparrow, which is not a sparrow and is now usually known as the dunnock, the stone curlew, which is not a curlew, and the water turkey, also called the anhinga, which is related to the cormorants, not to turkeys.

Coloration loosely relates a few birds to mammals as in fox sparrow and fox kestrel. The shape of the bill links Africa's whale-headed stork, also called the shoebill, to a whale; the enormous bill is as broad as it is long, giving the bird a rather ridiculous appearance. Again in Africa, there are mousebirds which have long tails and run and creep up vegetation. I used to watch them in my Kampala garden and always thought they looked more like mammals than birds. The bee hummingbird of Cuba is the smallest bird in the world, weighing only two grams, and certainly lives up to its name, not just because of its small size but also because of its mode of flight. The unique hoatzin of tropical America is often called the reptile bird because its young have claws on the first and second digits of the wings which, with the help of bill and feet, enables them to climb trees. One might imagine the turtle dove to be named after a turtle even though there is absolutely no resemblance or obvious association between the two animals. The name is, however, misleading, as it is derived from the French *tourterelle* which in turn is from the Latin *turtur*, an imitation of the bird's purring song. Equally misleading is brent goose which is named after an animal and not after Brent, a London suburb. Brent is derived from the Welsh *brenig* and the Breton *brennig*, meaning a limpet, but the connection is obscure except that both goose and mollusc are found on the sea-shore.

The list of borrowed names for insects is seemingly endless. Some have very strange origins as, for example, the word caterpillar which evidently comes from the Old French *chatepelose*, a hairy or downy cat. It must be assumed that the name was initially given to hairy caterpillars, in England still known as woolly bears or even as woolly boys. This interpretation makes nonsense of the word caterpillar for a kind of tractor that moves like a caterpillar. Other names are virtually impossible to comprehend and I have experienced much difficulty in trying to trace meanings. For a long time I was puzzled by the word cockroach, trying in despair to associate it with cock, a male bird, especially a chicken. Apparently it comes from the Spanish *cucaracha* which in 1624 was rendered as cacarootch by Captain John Smith who is quoted as saying: 'A certaine India Bug, called by the Spaniards a Cacarootch, the which creeping into Chests they eat and defile with their ill-scented dung.'

Behaviour has been extensively used to link insect names to the names of unrelated species. Tiny flea beetles have thickened thighs and jump with their hind legs rather like fleas. Cuckoo bees, also called homeless bees, resemble bumble bees in whose nest they lay their eggs. Worker bumble bees look after the developing larvae of cuckoo bees which upon becoming adult insects drive the queen bumble bee from the nest and take it over. This behaviour is reminiscent of the cuckoo which lays eggs in other birds' nests and does not take care of its own young. Antlions belong to the insect order Neuroptera; their larvae make pits in dry, sandy soil and bury themselves in the bottom of the pits where they wait with big, powerful jaws poised to seize ants which tumble into the pits.

Finally, there are dragon names, given to a variety of animals because of their resemblance to mythical dragons. Members of the insect order Odonata are collectively called dragonflies and, appropriately, the smaller, delicate species are known as damsel flies. A dragonet is a fish which is supposed to look like a little dragon; the males have a high, brightly coloured, spiny fin along the back and this, together with their ornamentally coloured bodies, accounts for the name. The famous flying dragons, lizards of the genus *Draco*, of South-East Asia have an extensible fold of skin on each side of the body which is supported by about half a dozen ribs. When extended these membranes look like wings and act as parachutes as the lizards leap from tree to tree.

−14−
Another way to immortality

IN AUGUST 1961 I crossed the Rio Grande at Brownsville in Texas and drove south into Mexico, taking the route to the province of Oaxaca and ending up at the narrow Isthmus of Tehuantepec which separates the Atlantic from the Pacific Ocean. High in the pine- and oak-clad mountains I saw familiar North American birds like pine siskins, crossbills and juncos, but when I descended towards the Atlantic coast I began to see unfamiliar species like toucans, parakeets and trogons, as well as gigantic, iridescent blue morpho butterflies. I had, in about two hours, crossed from one great zoological region, the Nearctic, to another, the Neotropical.

In Mexico, high mountains extend south like a pointed tongue, and on them most of the animals and plants are typical of North America. On both the Atlantic and the Pacific coastal lowlands most species are typical of South America. The difference between the two regions is for a zoologist both bewildering and exciting; bewildering because at first everything Neotropical is strange and identification a frustrating and difficult experience, although it is amazing how quickly one learns. I suppose a zoologist from the Neotropical region has the same feelings when he first enters the Nearctic.

The Nearctic zoological region comprises the whole of North America, including Greenland, and extends south on high ground to southern Mexico. The Neotropical comprises South and Central America, the lowlands of southern Mexico, and includes the West Indies, Cuba and associated islands. The animals of these two regions are very different, the Neotropical, which includes the vast rain forests of the Amazon basin, being far richer in species.

The world is divided into a further four zoological regions, making six in all. Europe, Asia north of the Himalayas and North Africa are called the Palaearctic region; Africa south of the Sahara the Ethiopian; India, South-East Asia east to the island of Bali the Oriental; Australia, New Guinea and all the islands west to Bali the Australian. New Zealand is usually included in the Australian but its animals are so distinct and different that some feel it should be regarded as a region in itself.

These zoological divisions of the world were first proposed in 1857 by the ornithologist P.L. Sclater on the basis of the distribution of birds, and were extended in 1876 by A.R. Wallace to take into account all other land animals. The six zoological regions, or realms as they are sometimes called, are now known as Wallace's regions; in the meantime Sclater's name has been forgotten.

Wallace was an excellent naturalist and an experienced traveller and explorer of the tropics. At about the same time as Charles Darwin was slowly pondering over his theory of evolution by natural selection, Wallace quite independently and much more quickly came up with essentially the same idea. The two men exchanged views and in 1858 jointly published an article outlining the theory. This was followed in 1859 by Darwin's *Origin of Species*, arguably the most significant book ever published in the history of biology, for which Wallace received little acknowledgement or recognition, despite his 1858 contribution. The theory of natural selection was proposed as a way of explaining how evolution occurred and how today's plants and animals have descended from more primitive forms. The theory depends on a series of propositions and observations, all of which can be detected in nature. The first is that plants and animals possess an enormous capacity to increase in numbers, but despite this numbers usually remain constant from year to year. A pair of blue tits produces ten or more young, a daisy plant produces thousands of seeds, while an oyster's offspring run into hundreds of thousands, each year. Yet we are not overpopulated with blue tits, daisies and oysters, nor with any other species of wild plant or animal. This means that death rates, especially among the young, are very high indeed: in some species over ninety per cent of those born are dead before they have a chance to reproduce.

Individual plants and animals of a species vary: rarely are two individuals exactly alike, and the majority of individual characteristics are inherited. Darwin and Wallace deduced from these observations that survivors are likely to differ in certain respects from those that die; in other words the likelihood of death (or survival) depends on an individual's genetic make-up. Those best adapted to their surroundings or environment are the survivors; if the environment changes a different set of individuals become the survivors. Hence this process of natural selection gives rise to change, or evolution. Evolution by natural selection is universally known as Darwinism. Wallace's contribution has been largely forgotten, but then he does have his zoological regions.

An alternative to Darwinism is Lamarckism, named after the French philosopher/biologist and intellectual jack-of-all-trades, Jean Baptiste Pierre Antoine de Monet, usually styled Chevalier de Lamarck, who in 1809 published his *Zoological Philosophy*. Lamarck's evolutionary theory suggests that individual plants and animals respond to specific needs and develop structures and behaviour to suit these needs. Such acquired characteristics (as they are called) are inherited by an individual's offspring which, in time, means that evolutionary change takes place gradually as one generation inherits what the previous one has acquired. The theory is no longer accepted, being replaced by the Darwinian explanation. Lamarck thought and wrote about many things and was often wrong, but he did invent the word biology and was the first to construct a family tree showing the evolutionary history of animals.

Wallace, Darwin and Lamarck are distinguished by

having their names firmly attached to their particular discoveries. This is another way to scientific immortality, perhaps even better than having a plant or animal named after you. Indeed all three men also have several species of plants and animals named after them, but it is for their scientific discoveries that their names are really remembered.

The father of the science of genetics was an Austrian abbot, Gregor Mendel, who by breeding plants, especially peas, and by making good use of his limited spare time, worked out the fundamental rules of inheritance upon which modern genetics is founded. We now speak of Mendelian inheritance while the study of heredity is often called Mendelism. Darwin did not known about Mendel's work, which was published in German in 1865, and so was never able to associate his theory of evolution by natural selection with the all-important rules of heredity, even though the necessary information was available. Both men were at fault: Darwin did not read German, and Mendel chose to publish his work in a journal that was hardly ever seen in England.

In 1798 Thomas Malthus, an English social economist, published his now famous *Essay on the Principle of Population* in which he argued that if the human population increased faster than the food supply there would be widespread famine. The Malthusian theory, as it became known, had a significant effect on Darwin's thinking because it brought home to him the enormously high death rates that must occur in all plants and animals. Armed with this information, Darwin felt better able to postulate that death and survival of individuals are not random events, and that those that die are less well adapted to the environment than those that survive, the essence of the theory of natural selection.

Animals compete and struggle with one another for scarce resources, especially for food, but also for places to hide from predators, to keep warm, and so on. In 1934, the Russian biologist, G.F. Gause, published a book called *The struggle for existence* in which he put forward the idea that no two species have identical ecological requirements. There are always differences, for example, in the way food is obtained, which are reflected in details of anatomy, and which reduce, to some extent, the struggle to survive. This important observation is known as Gause's thesis.

In 1840, a German chemist called Justus von Liebig published a book with the unlikely title of *Organic chemistry and its application to agriculture and physiology* in which he claimed, 'the crops on a field diminish or increase in exact proportion to the diminution or increase of the mineral substances conveyed ... in manure.' In other words, plant growth is restricted by those essential nutrients or trace elements in shortest supply, an idea now known as Liebig's law of the minimum, which is important not only in agriculture but for understanding the growth of plants in general.

These are just some of the better-known theories, laws, theses and -isms named after their discoverers or proponents. There are many more such as the less well-known Bergmann's rule which states that within a species of bird or mammal larger individuals occur in the colder and smaller individuals in warmer parts of the geographical range. This is to do with the need to conserve heat: a larger body means a relatively smaller surface area from which heat is lost. Related to Bergmann's is Allen's rule that in birds and mammals those parts of the body which stick out – tails, noses, ears, legs and beaks – are relatively shorter in colder than in warmer regions. This again is associated with heat conservation as heat is lost most rapidly from body parts that stick out. We all know that in cold weather it is the ears, nose, hands and feet that suffer most; Allen's rule may even explain why Eskimos have short, flat noses.

Then there are systems or ways of doing things. I have already mentioned (Chapter 2) Linnaeus's method of classifying the living world, now widely known as the Linnean system. Less well known is the Lincoln index, named after F.C. Lincoln, an American wildlife biologist who in the 1930s was studying ducks in order to find out how many could be shot without seriously depleting numbers. But first it was necessary to know the number of ducks available for shooting, information that could not be obtained by counting as ducks are too mobile, making it easy to count the same individuals again and again. Instead, Lincoln trapped the ducks and placed a numbered metal ring around the leg of each bird before releasing it. Subsequent captures of ducks then included both ringed and unringed individuals and by using simple arithmetic it was possible to estimate the true number present by comparing the proportion of ringed to un-ringed ducks. This method, which in fact had been used much earlier by fisheries' biologists interested in assessing fish stocks, is now used as a way of estimating the numbers of many kinds of animals, and is especially useful in situations where there is no chance of an accurate count of the individuals present.

Much more famous is the French scientist Louis Pasteur (1822–95) who discovered it was possible to kill organisms present in a product such as milk by heating at controlled temperatures that do not materially change the natural characteristics of the product. This process, called pasteurisation, is immensely important and has saved countless millions of people from the dangers of consuming food and drink containing harmful micro-organisms. Throughout the world there are now Pasteur Institutes concerned with research on all aspects of pasteurisation and related techniques for protecting food and drink from contamination.

The discovery of a new disease is rather like the discovery of an undescribed species of plant or animal and, as with new species, many diseases have been named after the discoverer. Huntingdon's chorea, an inherited disease causing involuntary muscle movement and a progressive worsening of the mental faculties, which starts early in middle age, was described in 1872 by George Huntingdon. It is said that he first saw the disease near New York while accompanying his father (a doctor) on medical visits. They came upon a mother and daughter bowing, twisting and grimacing in an idiotic-looking manner, and George, who was frightened by what he saw, later vowed to investigate the disease.

To cure a disease medicine is often required and throughout human history thousands of different medicines have been used for different ailments. Some of these medicines have been named after the doctors who first advocated their use, but although many of the ailments remain most of the medicines are no longer used – either because they have been replaced by better ones or because they have since been found to be ineffective. One of these latter is Dover's powder, a preparation of opium and ipecac (extracted from a low-growing South American shrub), once used to check spasms and to relieve pain by inducing sweating. It was named after Thomas Dover (1660–1742), an English physician who recommended its use.

As knowledge of treatments for diseases increased, so did the knowledge of human anatomy. New structures, many of them microscopic, were constantly being discovered which required naming, and so once again names of discoverers became immortalised, this time in parts of the body. One of the best commemorated is Marcello Malpighi, the seventeenth-century Italian doctor who was especially interested in the kidney and spleen. The filtering unit of the kidney of man and other vertebrates is called the Malpighian body, while the layer of skin containing melanin (the pigment causing dark coloration) is called the Malpighian layer. In addition the tiny excretory glands of insects are named Malpighian tubules. Another part of the kidney's anatomy is called the loop of Henle after Jacob Henle the nineteenth-century German anatomist. Henle was a Jew who had adopted Christianity and consequently became a victim of political persecution. He was arrested by the Prussian police and condemned for treason, but his scientific reputation saved him, as well as the intervention of Alexander von Humboldt after whom the ocean current is named.

In female mammals the structure which conveys eggs from the ovary to the uterus is called the Fallopian tube after Gabriello Fallopio, the sixteenth-century Italian anatomist who did so much to advance knowledge of the human sexual organs. Fallopio's youth was spent in poverty and he first entered the church before going on to study anatomy at Padua. He published a small but extremely useful work, *Observationes anatomicae*, before he died at the early age of forty. A Graafian follicle is a fluid-filled spherical chamber in a mammal's ovary containing a cell which eventually develops into an egg. It is named after Regnier de Graaf, the seventeenth-century Dutch physician who also distinguished himself by inventing the word ovary.

Hapsburg lip is the name given to a narrow, under-shot lower jaw and protruding underlip which results in the mouth being held half-open. It was present in several members of the Hapsburg dynasty and can be traced back to the fourteenth century, thanks to the many portraits that exist of members of that dynasty. It was present in, among others, Maria Theresa of Austria and Alfonso XII of Spain, but it is by no means confined to the Hapsburgs as it is found among people in many different parts of the world. The Hapsburgs have distinguished themselves in many ways but the name will probably persist in Hapsburg lip long after other distinctions are forgotten or relegated to the archives of history.

−15−
The companies of beasts and fowls

WHEN I WAS ABOUT sixteen years old, I used to catch the early Sunday morning electric train from Lewisham to Gravesend, and there wait for the little steam train that stopped at all stations to All Hallows on the North Kent marshes. On the platform at Gravesend I looked out for other bird-watchers, un-mistakable in dress and gumboots, each of them carrying a favourite or recommended brand of binoculars or telescope. Some of these men (only very rarely were there ladies present) were strangers from places as far away as Middlesex and North London; others were regulars, experienced men who knew the marshes and the birds. I always tried to associate with these experts, travel with them, and, unless they had other plans, spend a day bird-watching in their company, because then I would be sure to see more kinds of birds than if I remained alone. I remember trying to impress the experts with stories of what I had seen recently and with what I knew about birds and bird-watching. In retrospect, I must have been a rather tiresome youth, but these enthusiasts were kind and considerate and willing to help and listen.

Nearly always we left the train at the villages of Cliffe or High Halstow and then walked across the vast expanse of low-lying pasture to the sea-wall where, if the tide was out, the mud-flats stretched as far as the eye could see. There was one man in particular who was a real expert, not just because he could identify birds and knew where to go, but because he seemed to have eyes in the back of his head. He could use his long, heavy telescope with the precision of a soldier at rifle drill. We would be walking along the sea-wall when suddenly he would drop to the ground, cross one leg over the other, rest the telescope on his carefully positioned knees, and then, after what seemed to be no more than a cursory look through it, reel off what he could see: 'Twelve pintail out there, just by the tide-line. Twenty-three scoter flying at about two o'clock, just gone out of sight. Here come the waders, about a mile and a half out over the mud – about a thousand knot, with a few greys [grey plovers] among them.' Then one day, as we peered into the dim winter light, he dropped his telescope to the trail arms position, and announced: 'Charm of goldfinches just behind flying along the sea-wall. Probably continentals at this time of year.'

He had not actually seen the birds, just heard them

as they flew by, but it was this word 'charm' that was new to me. It somehow stuck in my mind and I started to use it whenever I saw a flock of goldfinches, explaining to anyone listening that this was the correct collective noun for a group of goldfinches. I went on to explain that it refers to their bright and attractive 'charming' coloration. Years later I discovered I was wrong.

The word dates from the fifteenth century and means a confusion of voices, not loud but quite unmistakable. Originally spelt 'cherme', then 'charme', the word has been used for a company of starlings, linnets (which have also been described as 'chirming'), children, and even angels. Richard Jefferies, writing in 1879, put it well: 'Thousands of starlings, the noise of whose calling to one another is indescribable . . . the country folk call it a "charm", meaning a noise made up of innumerable lesser sounds, each interfering with the other.' At about the same time, according to the *Oxford Dictionary*, Miss Jackson is supposed to have exclaimed: 'What a charm them children 'bin making i' school.'

So, a charm of goldfinches is heard, not seen, as is a murmuration of starlings. Collective nouns or nouns of assembly exist for many species of animals, but most are no longer in use. Evidently the Norman hunters and falconers were especially keen on them, and, according to some authorities, this was partly because they wanted to appear more knowledgeable than the crude and uninformed Saxons. The names rarely refer to the appearance of the animal; rather they offer an instant description of some idiosyncrasy of behaviour, real or imagined, often including, as in charm or murmuration, some aspect of sound or voice.

The best early source of collective nouns for animals is *The Boke of St Albans*, first published in 1486 and reputedly written, or at least compiled, by Dame (or Prioress) Juliana Berners. It is a fascinating book, full of all sorts of information, especially about hunting and hawking. Her suggested list of collective nouns is placed under the heading: 'The companies of beasts and fowls.'

Among them are others that draw attention to sound, including a chattering of choughs, but I think it likely that jackdaws are meant because the chough does not chatter and is unlikely to have been seen around St Albans, even in those days. Dame Juliana suggests duet for turtle doves and gaggle for geese if they are moving around noisily on the ground, or skein if they are flying over. She also suggests, rather unkindly, gaggle for a group of women, and, interestingly, bevy (the collective noun for quail) for a group of ladies. I suppose it is possible to imagine a similarity between a small group of rather plump quail and a group of ladies dressed in their fifteenth-century finery. Bevy is also used for roes and conies (rabbits), but not for partridges which occur in coveys. Grouse are also in coveys if they are all of the same brood, if not, pack is the correct word. Hounds come in packs, but another word is a cry, again emphasising sound, as does clamour for a group of rooks. Not all of these are listed in *The Boke*, but I suspect that the inspiration for inventing them stems from Dame Juliana.

Poets tell us that larks sing joyfully and bring a sense of well-being and a feeling of gladness to everyone in the right mood. Dame Juliana tells us that we should call a group of larks an exaltation, which sounds right as the word means rapturous delight, rejoicing, triumph, and 'a swelling of the heart'. It also means the act of leaping or springing up, which is what larks do when they are disturbed from a field.

Another word for a group of rooks is a building, which is a good description for their industrious communal nest-building at the rookery in early spring. Then there is a hill of ruffs, which is perhaps an attempt to describe the habit of males in the breeding season of congregating and forming a group on a slightly elevated piece of otherwise flat land. A desert of lapwings aptly describes the vastness of a winter flock numbering thousands and feeding over several acres of ploughed field – they seem to extend on and on, like a desert.

A fall of woodcock might refer to the appearance of immigrant birds in the fall, originally an English and more recently an American word for autumn, or it might be because they tend to turn up unexpectedly, as if they had fallen out of the sky. I once found a woodcock in a Kensington street where it looked lost and forlorn, and not long ago I disturbed one from a tiny backyard where it had been hiding in the compost heap. Both looked as if they had dropped or fallen in from somewhere, and both seemed totally out of place.

For herons, and sometimes for bitterns and cranes, the word siege (also written as sedge) is used to describe where these birds assemble and station themselves to wait for prey. In the days when herons were hunted with falcons it was vital for the falconer to know the whereabouts of the siege. As the *Oxford Dictionary* says: 'Having found the heron at siege, you must get you with your falcon up into some high place' and also 'a hern put from her siege . . . shall mount so high'.

The words herd and flock are, of course, in general use for groups of mammals and birds, respectively, and we can still speak of a herd of swans as if they are honorary mammals. Dame Juliana also suggests herd for curlews and wrens and, extending what I take to be a joke, for harlots. Neither herd nor flock suggest special traits or characteristics for mammals and birds in general, nor even, so far as I can see, for wrens, curlews or harlots in particular. However, most of Dame Juliana's collective nouns are reserved for particular species, and identify, or seek to identify, instantly recognisable characteristics. How about a shrewdness of apes, a business of ferrets, a watch of nightingales, a spring of teal, a labour of moles, a muster of peacocks, and a wisp of snipe? These seem to me to be just right, but others are perhaps less appropriate and certainly less fair. Nevertheless they still seem to capture something real or imaginary about the species in question.

A cur is a contemptible dog of low breed which is snappy and worthless; in bygone days the word might have been used for a poor man's watchdog, or even for a sheep-dog, and contrasted with the well-bred hunting dogs and hounds of the rich. Dame Juliana suggests cowardice for a group of curs, which seems unfair, and yet when you say a 'cowardice of curs' the slight alliteration sounds good. She also suggests an unkindness of ravens, which is really going too far, but then her entire list goes too far, well beyond the needs of hunter and falconer. It is when she suggests nouns of assembly for different sorts of people, especially for their varying professions, that she really has fun. Each word neatly picks out a particular idiosyncrasy or association of ideas, as in: an obedience of servants, a laughter of ostlers, a fighting of beggars, a melody of harpers, a poverty of pipers (I admit to being not too sure what this means), a skulk of thieves (also used for foxes and, interestingly, for friars), a pardon of vicars (exactly right, this one), a diligence of messengers, a superfluity of nuns, a dampening of jurors, and an incredibility of cuckolds. And, as if this last one is not enough, how about a noonpaciens (non-patience) of wives to be added to a gaggle of women and a bevy of ladies, mentioned earlier?

I am sure that Dame Juliana enjoyed listing these names for the companies of beasts and fowls, and I also like to think that nouns of assembly are still being invented. We all know that a group of thieves or thugs is called a gang, a word that has also been used (not by Dame Juliana) for a group of elk. In recent years the word gang has been used to describe a small assembly of politicians, as in the Gang of Four, whose political influence, however, was short-lived. Maybe one day gang will be the accepted word for a (small) group of politicians; it certainly sounds right, but I wish I could think of an apt word for a group of bird-watchers such as those I used to meet at Gravesend station.

Author's Book-List

ALLAN, P.B.M. *A Moth-hunter's gossip* P. Allan, 1937, 2nd edn Watkins and Doncaster, 1947.
One of the most erudite natural history books I have read. It has had an enormous influence on my development as a naturalist, and I feel sure I am not the only moth enthusiast to declare it a classic.

BLUNT, W. *The compleat naturalist: a life of Linnaeus*, Collins, 1971.
Biography of the first person to attempt to bring order into the nomenclature of plants and animals.

COTTLE, B. *Names*, Thames and Hudson, 1983.
A favourite of mine and invaluable to the searcher after truth about a wide variety of names.

GOTCH, A.F. *Mammals: their Latin names explained: a guide to animal classification*, Blandford Press, 1979.
GOTCH, A.F. *Birds: their Latin names explained*, Blandford Press, 1981.
Both books are useful guides to the Linnean system.

NORDENSKIÖLD, E. *The history of biology*, Kegan Paul, 1928.
Contains a wide range of information about names.

PARTRIDGE, E. *A Dictionary of slang and unconventional English*, Routledge and Kegan Paul, n.e. 1984.

PARTRIDGE, E. *The Penguin dictionary of historical slang*, edited by J. Simpson, Penguin Books, 1972, n.i. 1982.

Eric Partridge's book is accepted as the most comprehensive in its particular field. The Penguin edition is an abridged version.

POTTER, S. and SARGENT, L. *Pedigree: essays on the etymology of words from nature*, Collins, 1973.

An excellent volume, to be read and reread, especially good to dip into when one is in the mood for digression.

PRIOR, R.C.A. *On the popular names of British plants, being an explanation of the origin and meaning of the names of our indigenous and most commonly cultivated species*, F. Norgate, 3rd edn 1879.

A mine of information combined with delectable nineteenth-century pendantry, prudery and prejudice.

SWAINSON, C. *Provincial names and folk-lore of British birds*, English Dialect Society, 1885.

The most useful volume in my collection of books about local and dialect names.

THOMAS, K. *Man and the natural world: changing attitudes in England, 1500–1800*, Allen Lane, 1983.

Gives a useful historical perspective to the relationship of man with his environment.

Index

Admiral (butterflies)
76
red 37, 76
Aircraft, named from
animals 94–5
Albin's Hampstead
eye 99
Alchymist (moth) 65,
79
Alexanders 102
Allen's rule 125–6
Angler (fish) 79
Anomalous (moth) 65
Ant, driver 79
Pharaoh's 77
tailor 72, 73
Ant-lion 118
Apollo (butterfly) 37
Apostle-bird 75
Archer (fish) 79
Arches, Kent black 100
Kew 100
Arrow worm 90
Arum maculatum 46
Assassin bugs 76

Babbler,
arrow-marked 44
Backswimmer (bug)
78
Basket (star) 87
Bat, bulldog 116
mastiff 116
Bed-bug 86
Bee, carpenter 77
cuckoo 118
mason or potter
77–8
Beetle, bombardier 75
cardinal 73
carpet 86
Colorado 105
flea 118
sailor 72, 73
timberman 78
Bell-bird 89
-magpie 88

Bergmann's rule 125
Bevy, of coneys 133
of ladies 133
of quail 133
of roes 133
Birds, named after
places 103–4
Birds, on American
postage stamps
104–5
Bishop (bird) 73–4
Bishop's mitre (bug)
73
Blackbird 64
Blenny, sabre-toothed
89
Blue, adonis 36
mazarine 29
Blue heart playboy
(butterfly) 80
Bobolink 44
Bowfin (fish) 90
Box-fish 87
-turtle 87
Brown, Arran 99
evening 36
meadow 36
Buddleia 27–8
Building, of rooks 134
Bunting, Lapland 103
Burnet, New Forest
100
Scotch 100
Burnet companion 67
Bush-crow,
Stresemann's 25
Business, of ferrets 135
Butcher-bird 78
Buttercup 13, 18
Butterfly clam 109–10,
115
fish 115
By-the-wind sailor
(jellyfish) 75

Cabbage (moth) 66
Camberwell beauty 98

Cambridge vagrant 76
Canary 107
Canoe-shell 88
Cardinal (beetle) 73
(birds) 73
Carp, mirror 88
Carpet (beetle) 86
(moth) 86
(shark) 86
(shell) 86
Arran 100
autumn green 66
Devon 100
garden 9
Cars and lorries,
named from
animals 95
Catbird 41
Caterpillar 117–18
Celandine, lesser 54
Charaxes 33–6, 36–7, 37
Charm, of goldfinches
131–2
Chattering, of choughs
133
Chickadee 43
Carolina 104
Chicken 60
Pharaoh's 77
Chief (butterfly) 74
Chiff-chaff 41, 42–3
Chimney sweeper
(moth) 79
Chimpanzee 60–2
Chough 41, 134
Chuck-will's-widow
44
Clam 85
butterfly 109–10,
115
Clamour, of rooks 133
Cleaner (fish) 77,
89–90
Clearwing, Welsh 100
Cockroach 97, 118
Comb-oyster 88
Conformist (moth) 73

Copper (butterflies) 76
     large 76
Coronet, Barrett's
     marbled 28
Cosmopolitan 100
Cousin german (moth)
     65–66
Covey, of grouse 133
     of partridges 133
Cowardice, of curs 135
Cowslip 49
Crab, hermit 80
     spider 110
Crowbar 93
Cry, of hounds 133
Cuckoo 39, 41, 51, 63
     didric 44
Cuckoo-names 46–7,
     50–1
Cuckoo spit 51
Curlew 41
     stone 116
Cushion-star 87
Cutlass-fish 90
Cuttlefish 111

Dafila (girl's name) 92
Dampening, of jurors
     136
Damsel fly 80, 119
Dandelion 49
Dart, Gregson's 28
Darwinism 123
December (moth) 66
Deer, mouse 110
     Père David's 28
Delicate (moth) 65
Demoiselle (crane) 80
Desert, of lapwings
     134
Diligence, of
     messengers 136
Dog 14, 107
Donax 109
Dove, mourning 41
     turtle 117, 133
Dover's powder 127–8
Dragon, flying 119
     Komodo 106–7
Dragonet 119
Dragonfly 118
Drill (snail) 85
Drinker (moth) 66
Dropwort, water 20
Drum fish 89

Duck, long-tailed 41
     mandarin 80
     sawbill 83–4
     shoveller 77, 85–6
Duck-billed platypus
     116
Duet, of turtle doves
     133
Dunnock 21, 92, 116

Eagle, Bonelli's 24
     wedge-tailed 85
Emerald, Essex 100
     Sussex 100
Emperors and
     empresses 77
Exaltation, of larks
     133–4
Exile (moth) 65
Exocet 96
Eyebright 54

Falcon, cuckoo 116
     Eleonora's 26
Fall, of woodcock 134
Fallopian tube 128–9
Fieldfare 42
Fighting, of beggars
     136
Figwort 53–4
Filefish 85
Fish, named after
     mammals 115–6
Flea 14
Fleabane, Canadian
     100–1
Footman (moth) 77
Forester (moth) 78
Forkbeard (fish) 86
Friar (bird) 73
     (butterfly) 73, 74
Fritillary (flower) 60
     Glanville (butterfly)
     26–7
Frog fish 115
Fruit fly 13, 18

Gaggle, of geese 133
     of women 133
Gallinule, Florida 104
Gang, of politicians
     136
     of thieves or thugs
     136
Gardenia 30

Gatekeeper (butterfly)
     76
Gause's thesis 125
Gem, Cumberland 100
     Tunbridge Wells 71
Geometrician (moth)
     78
Geotrupes stercorarius 50
Ghost (moth) 69
Gipsy (moth) 79–80
Girls' names from
     flowers 91–2
Glory, Ealing 100
     Kentish 100
Goat (moth) 68
Go-away bird 44
Goldeneye, Barrow's
     22
Goldfinch, European
     108, 131–2
Good King Henry
     19–20
Goose, barnacle 100
     brent 117
     Canada 97
Goose barnacle 110
Graafian follicle 129
Grass-of-Parnassus
     102
Grebe, great-crested
     50
     little 50
Grenadier (fish) 75
Grey dagger 9
Grindle 90
Guinea fowl 60
Gull, Bonaparte's 24
     California 104
     laughing 40
     Sabine's 25

Hadada 44
Hairsteak, white-letter
     99
Hammerhead (shark)
     84
Hammerkop (bird) 84
Handsaw (heron) 82–
     3
Hapsburg lip 129
Harlequin, animals
     so-named 115
Harrier 94
     Montagu's 23, 26
Hawk (bird) 82–3

Hawk (moth), death's
  head 69–70
  privet 9
Hawkweed 54
Helmet-shrike 90
Herald (moth) 66
Herb paris 102
Herd, of curlews 135
  of harlots 135
  of swans 135
  of wrens 135
Hermit (crab) 80
  (humming bird) 80
  (thrush) 80
Heron 18, 83, 134
  boat-billed 88
  Lousiana 104
Herpes 81
Herpetology 81
Highflyer (moth) 66
Hill, of ruffs 134
Hoatzin 117
Homo erectus 15
  neanderthalensis 15
  sapiens 15, 62
Hoopoe 42
Humboldt current 128
Hummingbird,
  bearded
    mountaineer 80
  bee 117
  fairy 80
  hermit 80
  hooded visor-bearer
    90
  sword-tailed 90
  white-tipped
    sicklebill 85
Huntingdon's chorea
  127

Ichneumonidae 31, 57,
  81
Incredibility, of
  cuckolds 136

Jack-names 42
Jackass, laughing 61
Jack-by-the-hedge 49
Jackdaw 42
Jay 18–19
  Florida 104
  Siberian 103
Joker (butterfly) 80
Junco, Oregon 104

Katydid 39, 41
Kestrel 42
  fox 117
Killdeer 43
Kings and queens
  76–7
Kite, Mississippi 104
Kittiwake 41
Knifefish 86

Labour, of moles 135
Lackey (moth) 65
Ladybird 51
  two-spot 18
Lady's bedstraw 51, 52
Lady's slipper 51, 52
Lady's smock 50–1
Lady's tresses 51
Lamarckism 123
Lamp-shells 86
Lanternfish 86
Laughter, of ostlers
  136
Layman (butterfly) 74
Liebig's law of the
  minimum 125
Lily, Snowdon 102
Lincoln index 126
Linnean system 14–17,
  126
Linnet 19
Lobster (moth) 67
Loop of Henle 128
Loosestrife, purple 47
Lords and ladies 46–7,
  48, 55
Lungwort 53
Lyrebird 88

Magpie 93, 112
  animals named after
    it 112–3, 113–4
  dialect names 113
Maiden's blush 65
Malpighian body 128
  layer 128
  tubules 128
Malthusian theory 124
Manchester treble-bar
  100
Mandarin (duck) 80
  (orange) 80
Man-o'-war bird 89
Mantids 75
March (moth) 66

Melody, of harpers 136
Mendelism 124
Miller (moth) 69
Minor, Haworth's 29
Missiles, named from
  animals 96
Mocha, Blair's 29
Mocking-bird 41
Mole 96, 135
Monarch (butterflies)
  74
Mongoose, Egyptian
  81
Monk (butterfly) 73,
  74
Monkey, howler 39
  spider 116
Monkey-wrench 93
Mosquito 19
Mother shipton
  (moth) 67
Mouse, deer 110
Mousebird 117
Murmuration, of
  starlings 132
Muster, of peacocks
  135

Necklace shell 88
Needlefish 88
Needle-whelk 88
Nightjar 26, 41
Nightshade, deadly
  103
Noddy, brown 62
Non-conformist
  (moth) 73
Noonpaciens, of wives
  136
November (moth) 66
Novice (butterfly) 74
Nuthatch, Corsican
  103
Nymphalidae 36–7

Oak eggar (moth)
  67–8
Oarfish 88
Obedience, of servants
  135
Oil-bird 88
Old lady (moth) 9, 80
Old man's beard 52
Orchid, early purple
  47–8

Organ-pipe cactus 88
    coral 88
Oriole, Baltimore 104
    golden 64
Oven-bird 86
Owenus minor 31
Owl, little 62
    Tengmalm's 23
    Ural 103
Oxslip 49

Pack, of grouse 133
    of hounds 133
Paddlefish 88
Painted lady 10–11, 76
    American 105
Pansy 45–6
Pardon, of vicars 136
Parrot-bill,
    ashy-headed 58
Parson-bird 74–5
Partridge, Mrs
    Hodgson's 59
    red-legged 53
Pasteurisation 127
Peacock 60, 135
Pepper-pot (shell) 86
Petrel, Leach's 23
    Wilson's 24
Petroleum fly 88
Pewee 43
Phoebe 43
    Say's 24
Phoenix (moth) 65
Pied piper (butterfly)
    80
Pilot fish 76
Plane, London 101
Plover, blacksmith 79
    Kentish 103
    little ringed 61
    Wilson's 24
Plusia, Dewick's 29
Policeman (butterfly)
    76
Poorwill 44
Pope (fish) 74
Portland moth 100
Portuguese
    man-o'-war (fish)
    89
    (jellyfish) 89
Poverty, of pipers 136
Precis butterflies 36–7,
    76–7

Prinia (girl's name) 92
Prunella (girl's name)
    92
Pseudacraea 59
Pseudoneptis 59
Pseudopontia 58, 59, 60
Puffin 75
Pug, Edinburgh 100

Quail, button 88
Quaker (moth) 73
Queen (butterfly) 74

Rabbit 85
    snowshoe 87
Ragged robin 54–5
Ragwort, Oxford 101
Rail, Virginia 104
Rannoch brindled
    beauty 100
    looper 100
    sprawler 100
Rat, brown 61
    Pharoah's 77
Rattlesnake 89
Razorshell 87
Red admiral 37, 76
Redwing, Owen's 31
Rhubarb 103
Rifleman (bird) 75
Rivulet, Blomer's 29
Robber flies 76
Robin, American
    105–6
Rock-fowl,
    bare-headed 59
Rustic, Eversmann's
    29
    Vine's 29

Sailor (beetle) 72, 73
Satyridae 36
Saw-bill 83–4
    -fish 83
    -fly 84
    -shark 83
Saxifrage 53
Saxon (moth) 65
Scabious 53
Scotch annulet 100
    argus 56–7, 99
    burnet 100
Sea-urchin 80
    hat-pin 87
Secretary (bird) 78–79

Self-heal 21, 92
Setaceous hebrew
    character 65
Shark 83, 84
    carpet 86
    hammerhead 84
    nurse 78
    saw- 83
Shearwater, Cory's 23
    Manx 103–4
Shellfish 110–11
Shepherd's purse 52
Shieldbug 90
Shoulder-knot, Blair's
    27–8
Shoveller (duck) 77,
    85–6
Shrewdness, of apes
    135
Shrimp, fairy 80
Sicklebill (shrike) 85
    white-tipped
    (hummingbird)
    85
Siege (or sedge), of
    bitterns 134
    of cranes 134
    of herons 134
Silverfish 111
Skein, of geese 133
Skipper, Essex 98
    grizzled 76
    Lulworth 98
Skua, Arctic 50
Skulk, of foxes 136
    of friars 136
    of thieves 136
Snipe, jack 42
    painted 116
    Wilson's 24
Snout, Paignton 99
Soldier fly 72, 73
Sparrow, English 106
    fox 117
    hedge 116
    Ipswich 104
    Savannah 104
Spider 93
    crab 110
Spinach 103
Spinster 93
Spittlebug, meadow 51
Spoonbill 86
Sprawler (moth) 65
    Rannoch 100

Spring, of teal 135
Spring usher (moth) 66
Squirrel, American grey 105
Star of Bethlehem 52, 102
Star-gazer (fish) 80
Starling, European 106, 132
Stint, Temminck's 23
Stork, adjutant 75
  bishop 73
  shoe-billed 86, 117
  whale-headed 86, 117
Stranger (moth) 65
Superfluity, of nuns 136
Surgeon (fish) 78
Swan, Bewick's 23, 25
  whistling 40
  whooper 40–41
Swede 102
Swift 42
Swiftlet, white-bellied 61–2
Sword-tail (fish) 90

Tanks and armoured cars, named from animals 94
Tern, Sandwich 103
Thorn, August (moth) 66
  September (moth) 66
Thrush, hermit 81
  rock 18
  Siberian 103
  White's 23
Tiger, cream-spot 70
  Jersey 100
Tinker bird 72–3
  yellow-fronted 72–3
Tit, long-tailed 42
  marsh 19
  Siberian 103
  willow 19
Toad, midwife 78
  natterjack 42
  spadefoot 85

Toadstool 19
Tortoise, tent- 88
Towhee 43–4
True-lover's knot 65
Trumpeter (bird) 78
Turkey 60
  water 116
Turnip (moth) 66

Umber, Cumbrian 100
Umbrella-bird 87
  -tree 87
Uncertain (moth) 65
Underwing, yellow 9
Unkindness, of ravens 135

Veery 43
Vervain 55
Violet, dame's 102
Viper's bugloss (moth) 66
Vireo, Philadelphia 104
Vole, water 60
Vulture, Egyptian 103
  turkey 59

Wainscot, Blair's 29
  Brighton 99, 100
  Devonshire 100
  fen 29
  Fenn's 29
  Mathew's 29
  Webb's 29
Wallace's regions 122
Wallflower 52
Warbler, Bonelli's 24
  Connecticut 104
  Dartford 103
  grasshopper 40
  Kentucky 104
  Mrs Moreau's 25
  Nashville 104
  Tennessee 104
  Wilson's 24
  Winifred's 25
Wasps, parasitic 31, 81
Watch, of nightingales 135
Water betony (moth) 66

Water-thrush, Louisiana 104
Watering-pot (shell) 86
Wave, Isle of Wight 100
  Lewes 99
  Portland riband 100
  Weaver's 29–30
Waxwing, Bohemian 80
Weapons, named from animals 95–6
Weasel, stripe-bellied 61
Weaver-bird 78
Wheatear 21, 49, 50
Whippoorwill 38, 44
White, Bath 58, 98
  cabbage 53
  green-veined 15
  large 15
  moth-like 58
  small 15, 16
Widow, black (spider) 80
  crazy (bird) 81
Willet 43
Willowherb, rosebay 52–4
Wisp, of snipe 135
Wood-hewer (bird) 78
Woodlark 41–2
Woodlice 48
Woodpecker, green 41, 42, 64
Woolly bear 118
Workers (ants, bees, wasps, termites) 79
'Worms' 111–2
Wren, Carolina 104, 106
  winter 106
Wryneck 63, 81

Yellow hammer 84–5
Yorkshire fog 102

Zebra, animals named after it 114–15